Content-Based Audio Classification and Retrieval for Audiovisual Data Parsing

THE KLUWER INTERNATIONAL SERIES
IN ENGINEERING AND COMPUTER SCIENCE

CONTENT-BASED AUDIO CLASSIFICATION AND RETRIEVAL FOR AUDIOVISUAL DATA PARSING

TONG ZHANG AND C.-C. JAY KUO
Integrated Media Systems Center and
Department of Electrical Engineering - Systems
University of Southern California
Los Angeles, CA 90089-2564, USA

Kluwer Academic Publishers
Boston/Dordrecht/London

Distributors for North, Central and South America:
Kluwer Academic Publishers
101 Philip Drive
Assinippi Park
Norwell, Massachusetts 02061 USA
Telephone (781) 871-6600
Fax (781) 871-6528
E-Mail <kluwer@wkap.com>

Distributors for all other countries:
Kluwer Academic Publishers Group
Distribution Centre
Post Office Box 322
3300 AH Dordrecht, THE NETHERLANDS
Telephone 31 78 6392 392
Fax 31 78 6546 474
E-Mail services@wkap.nl>

Electronic Services <http://www.wkap.nl>

Library of Congress Cataloging-in-Publication

A C.I.P. Catalogue record for this book is available from the Library of Congress.

ISBN 978-1-4419-4878-6

Printed on acid-free paper.

Printed in the United States of America

Contents

List of Figures

List of Tables

Preface

The automatic segmentation, indexing and retrieval of audiovisual data has been an extremely hot research topic recently which can be a significant technique in many areas such as the delivery of video segments for professional media production, storage and retrieval of video databases, video-on-demand, user agent driven media selection and filtering, semantic multicast backbone video, as well as in educational applications and surveillance applications. While previous research on this topic primarily focuses on the pictorial part, significant clues contained in the accompanying audio flow are often ignored. However, more and more people have realized that a fully functional system for video content parsing can be achieved more successfully through a proper combination of audio and visual information.

In this research monograph, an overview of existing approaches for audio and video content analysis is given first, then a system is proposed for audiovisual data segmentation, indexing and retrieval based on multimodal media content analysis. The purpose is to generate meta-data for video sequences for information filtering and retrieving. The audiovisual stream is de-multiplexed into different media types such as audio, image and caption. An index table is generated for each video clip by combining results from content analysis of these diverse media types. Structures for different video types are described, and models are built for each video type individually. This general modeling and structuring of video content parsing is very unique. It achieves more functions than existing approaches which normally adopt a single model with focus on the pictorial information alone.

Audio, which includes speech, music, and various kinds of environmental sounds, is an important type of media, and also a significant part of audiovisual data. Compared to research done on content-based image and video database management very little work has been done on the audio part of the multimedia stream. Therefore, a large part of this monograph is devoted to the content-based management of audio data, and a hierarchical system consisting of three stages is developed. In the first stage, the task of on-line segmentation and classification of accompanying audio signals in audiovisual data is accomplished. Segment boundaries are first detected by locating abrupt changes in short-time audio features, and each segment is then classified to be one of the basic audio types including speech, music, environmental sound, song, speech with music background, silence, etc. It should be noticed that besides commonly studied audio types such as speech and music in existing work, we have taken into account hybrid types of sound which contain more than one kind of audio component like speech or environmental sound with music in the background, because their distinction is often important in characterizing audiovisual segments. The proposed procedure is generic and model free which can be easily applied, as the first step processing of digital audiovisual data, to almost any content-based audiovisual material management system. Experimental results have shown that audio boundaries are precisely set, and an accurate classification rate higher than 90% is achieved.

In the second stage, further classification is conducted within each basic audio type. In this monograph, we especially put emphasis on the distinction of environmental sounds which are often ignored in previous work. Environmental sounds, including sound effects, are an important ingredient in audiovisual recordings, and their analysis is essential in many applications such as the post-processing of films and the retrieval of video segments with special sound features. We investigate the classification of sound effects into semantic classes such as applause, footstep, explosion and raining, based on the time-frequency analysis of audio signals and the use of the Hidden Markov Model (HMM) as the classifier. Experimental results show that an accuracy rate of 86% is obtained. Finally, in the third stage, an audio retrieval approach is proposed on top of the audio archiving schemes. A query-by-example retrieval scheme for sound effects has been proved to be very effective and efficient. That is, with a given query sound, similar sounds in an database can be automatically retrieved. This provides an important tool for searching specific audio clips within audio or video databases.

In this research, we have also worked on content analysis of the visual part of audiovisual data. First, an efficient and robust method is developed for shot change detection in image sequences. This new method is derived from the twin-comparison algorithm with a new ingredient, i.e. the histogram difference of the Y- and V- components are incorporated. It is shown that the proposed method achieves for both the sensitivity rate and the recall rate at around 95% with various kinds of test video. Second, a scheme is proposed for the adaptive keyframe extraction based on histogram comparison. It is demonstrated by experiments that it can generate keyframes that properly represent the content of a shot.

Both theoretical analysis and implementational issues are described in detail in this monograph. For example, the readers will find the procedures for estimating fundamental frequency and spectral peak tracks, the process for training a hidden Markov model and the rules for distinguishing music components from non-music components. A large number of examples are also included (especially in the form of figure) to show properties of audio features, different types of video, various situations for shot change detection, analyses of experimental results and so on. We hope by reading this monograph, the readers will get a clear and vivid view of the characteristics of audiovisual data, as well as methods to analyze these data using digital signal processing and pattern recognition techniques.

TONG ZHANG

Acknowledgments

Some experimental results and illustrations shown in this book are based on the video part of MPEG-7 test data. They are used here for the purpose of scientific research only.

To my parents and
Jerry.

I
INTRODUCTION

Chapter 1

INTRODUCTION

1. SIGNIFICANCE OF PROPOSED RESEARCH

1.1 VIDEO SEGMENTATION AND ANNOTATION

The automatic segmentation, indexing, and retrieval of audiovisual data has important applications in professional media production, audio-visual archive management, video-on-demand, user agent driven media selection and filtering, semantic multicast backbone video, education, surveillance, and so on. For example, television and film archives store a vast amount of audiovisual material. If these data are properly indexed at the segment level, it will provide a great help for studios in delivering video segments for editing an archive-based video, a documentary, or an advertisement video clip. In audiovisual libraries or the family entertainment applications, it will be convenient to users if they are able to retrieve and watch the video segments of their interest. As the volume of the available material becomes huge, manual segmentation and indexing is impossible. The most proper way to achieve segmentation and indexing is through computer processing based on content analysis of audiovisual data.

Current approaches for video segmentation and indexing are mostly focused on visual information (such as color histogram differences, motion vectors, and keyframes) [1], [2], thus neglecting the content of the accompanying audio signal and closed caption. In fact, there is an important portion of information contained in the continuous flow of audio data which may often represent the theme in a simpler fashion than the pictorial part. For instance, all video scenes of gun fight should include the sound of shooting or explosion, while the image content may vary significantly from one video clip to another. In the beginning of the movie "Washington Square", there is a segment which is of several minutes long, showing the buildings, streets, and people of the neighborhood. There are many quite different shots involved, but the continuous accompanying music indicates that they are actually within one segment. Moreover, the speech information contained in audio signals is usually critical in identifying the theme of the video segment. Normally, by only listening to the dialog in a segment, it is enough for us to understand what it is about. However, we may be easily lost by watching the pictures only. Thus, it is believed that a fully functional system for video content parsing can only be achieved based on the integration of diverse media (audio, visual, and textual) components

1.2 AUDIO AND VISUAL CONTENT ANALYSIS

Audio, which includes voice, music, and various kinds of environmental sounds, is an important type of media, and also a significant part of video. Compared to research done on content-based image and video database management, very little work has been done on the audio part of the multimedia stream. However, since there are more and more digital audio databases in place these days, people begin to realize the importance of effective management for audio databases relying on audio content analysis. In content-based audio classification and retrieval, audio data are automatically archived into different audio types and semantic classes, and we can search for a particular sound or a class of sound in a large audio database electronically based on the content analysis of audio signals. There is a wide range of applications of this technique in the following domains.

1)Entertainment. In the film industry, it will be extremely helpful to be able to search sound effects automatically from a very large audio database, which contains sounds of explosion, windstorm, earthquake, animals, and so on. It is especially important in animation film making where a whole environment has to be created in one shot, which means dealing with huge databases. There are also applications in producing TV and radio programs. For example, editing a concert recording would be aided by sound classes of audience applause, solo instruments, loud and soft ensemble playing, and other typical sound features of concerts; and these sounds can be provided by the audio retrieval algorithms.

2)Audio Archive Management. Great convenience will be provided for material searching and browsing in audio libraries with automatic audio classification and retrieval. There are some big audio libraries around the world. For example, the NFB of Canada's sound effects library currently includes 1200 hours of analog tape material, representing some 40000 sound effects. Access is time-consuming, requiring editors to make selections from a catalogue, manually locate and handle tapes for auditioning purposes. There is an urgent demand to search for effects electronically. While using keywords for retrieving the effects is a method, it is still time and labor consuming for indexing, and has the problem of lacking objectiveness and consistency, especially for those features of sound which are hard to describe. Therefore, the best way is to index and search the sound effects by content-based audio classification and retrieval.

In large archives of raw audio, it is useful to have some automatic segmentation and indexing for the raw recordings. For example, in arranging the raw recordings of performances, meetings, or panel discussions, segments of silence or irrelevant environmental sounds (including noise) may be discarded, while speech, music and other environmental sounds can be classified into the corresponding archives.

Much more flexibilities will be provided for indexing and browsing an audio database with audio content analysis. For instance, when browsing an audio database, the user would be supplied with a menu from which he can choose different combinations of features of the sounds he desires. These features can be obtained by content-based analysis and indexing of the audio data.

3)Commercial Musical Usage. The Karaoke industry is extremely large and popular. A much more friendly interface will result if the client is allowed to hum a few memorable bars of the requested tune, and let the computer to retrieve the song. In music stores or on-line shopping, clients may hum approximate renditions of the song they seek from a kiosk or from the comfort of their own home. Alternatively, they may seek out music with similar features to those they already know. From there, they may listen to appropriate samples and choose to buy the products on the spot.

4)Surveillance. Many offices are already equipped with computers that have built-in audio input devices, which could be used to listen for the sounds of people, glass breaking, and so on. Such surveillance could also be used to detect sounds associated with criminal activities in areas which have the needs, and trigger an action if some event occurs.

5)Audio Coding. In MPEG-4 Parametric Audio Coding, speech and broadband audio (such as music) signals are coded with different schemes in order to get the best overall performances [3]. Therefore, audio classification algorithm is needed to separate the speech signal from broadband audio in the audio bit stream to compress them respectively. Audio classification techniques can also be used in the voice activity detection (VAD) part of variable rate speech coding[4], [5]. For this application, we would like to distinguish active speech segments from pauses, when the speaker is silent and only background acoustical noise is present. An effective VAD algorithm is critical in achieving low average rates without degrading speech quality in wireless communication.

6)Others. In medicine, audio content analysis can be used to detect certain kinds of sounds for the diagnosis of certain kinds of diseases, e.g. to monitor the cries of babies, the heart sound, and so on.

Audio classification is also useful in monitoring TV and radio channels. For example, it will be possible to automatically change channel when disliking contents come, such as commercials or talking programs. In surveillance which monitors several radio channels, it can be used to effectively ferret out intelligence by ignoring channels playing music. In film rating, audio content analysis can be used to detect sounds related to violence in films (shots, explosions, cries, etc.) and give a rate of violence of the film.

In this work, visual content analysis consists of the segmentation of image sequences, the extraction of keyframes, and the low-level to semantic level feature analysis of shots and keyframes in a video stream. It has been studied as the major technique in detecting, isolating, and representing meaningful segments in video sources. A segment is normally defined as a single, uninterrupted camera shot [6]. This simplifies the partitioning task to detecting boundaries between consecutive camera shots. Besides its obvious significance in video indexing, image sequence segmentation can be used in video editing for scene change detection. It is also important in motion compensation for video compression, because motion vectors should be computed within segments rather than across segment boundaries. Content-based classification and retrieval of images has received a lot of attention in the past several years, which has applications in video and image database management (indexing and browsing), video conferencing, on-line shopping, electronic photo album, medical imaging diagnoses, etc.

1.3 MPEG-7 STANDARD DEVELOPMENT

In October 1998, MPEG started a new work item, namely, the "Multimedia Content Description Interface", or in short "MPEG-7", which aims to specify a standard set of descriptors and description schemes that can be used in describing various types of multimedia information [7], [8]. This description shall be associated with the content itself, to allow fast and efficient search for material of a user's interest. The major objective of MPEG-7 is to make audiovisual material "as searchable as text". Examples may be to search for "twenty minutes of video according to my preferences of today", or the scene of "King Lear congratulates his assistants on the night after the battle".

Audiovisual material that has MPEG-7 data associated with it, can be indexed and searched for. This "material" may include: still pictures, graphics, 3D models, audio, speech, video, and information about how these elements are combined in a multimedia presentation. Therefore, audio/visual content description will be important ingredients of MPEG-7, and audio/visual feature extraction, audio/video classification, and content-based audiovisual data retrieval techniques will be directly related to MPEG-7.

The scheme of using MPEG-7 in a multimedia database system is shown in Figure 1.1, where the description of materials in a multimedia database is generated through the MPEG-7 mechanism (contents within the two rectangular frames with dotted margins are normative parts of the MPEG-7 standard). This description is encoded, and transmitted or stored. Then, at the user's end, the description is decoded, and through the search/query engine or the filtering agents, those materials satisfying the user's requests will be retrieved and presented to the user.

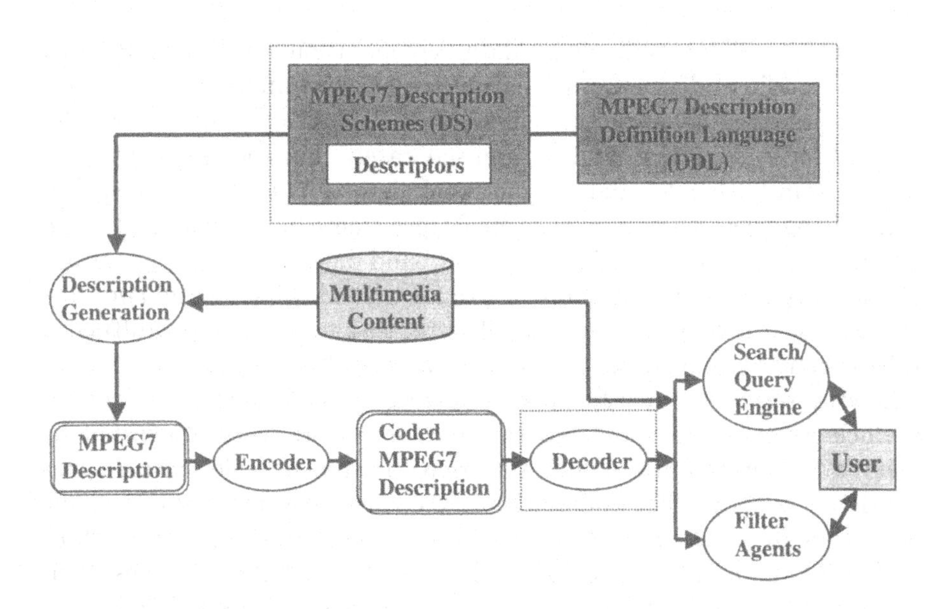

Figure 1.1. An abstract representation of possible applications using MPEG-7.

2. REVIEW OF PREVIOUS WORK

2.1 WORK ON VIDEO INDEXING AND RETRIEVAL

Previous work on video segmentation and annotation are primarily focused on the visual information. A common framework is to detect video shot changes using histogram differences and motion vectors, and extract keyframes to represent each video shot [6], [9]. However, this visual-based processing often leads to a far too fine segmentation of the audiovisual sequence with respect to its semantic meanings. For example, in one video clip of news program about a congress meeting, there are a dozen of shots including those of the anchorperson, of the speakers, of the audience, and several broad views of the hall. Based on the visual information, these shots will be indexed separately. However, according to the audio information, the continuous speech of the anchorperson indicates that they are actually within one news item. Furthermore, there are often significant clues contained in the audio data, e.g. the sounds of song performances in a music show video, the sounds of whistle and applause in a sports video, and sounds of conversation and music in feature movies, which provide enormous help in segmenting and annotating the video content.

A new trend for audiovisual data segmentation and indexing is to combine audio and visual information under one framework. This idea was examined in some recent papers. In [10], break detectors were developed for audio, color, and motion separately. For shot segmentation, results from both color and motion break detection were combined. For scene change detection, one looks for frames where both visual and audio breaks were detected. In [11], audio subband data and color histograms of one video segment were combined to form a "Multiject", and two variations of the hidden Markov model were used to index the Multijects. Experimental results for detecting the events of "explosion" and "waterfall" were reported. In [12], the color histogram differences, the cepstral coefficients of audio data, and the motion vectors were integrated by using a hidden Markov model approach to segment video into regions consisting of shots, shot boundaries, and camera movement within shots. Even though it has been demonstrated that these methods can combine the visual and audio information to handle some specific scenarios, their applicability to a generic video type is still yet to be proved. Meanwhile, more delicate audio feature extraction methods should be considered to provide more robust segmentation results.

There are also several systems or structures proposed, which consider the description of audiovisual contents by using more than one type of media [13], [14]. Some interesting high-level concepts have been presented, including object-oriented description, multi-level description, multiple modality, metadata dictionary, table of content, analytical index, etc. However, low- to mid-level implementations of audio, visual and textual content analysis and their integration are still lacking in these systems. Proper algorithms have to be developed to support these high-level ideas.

2.2 WORK ON AUDIO CONTENT ANALYSIS

Existing research on content-based audio data management is still quite limited, which can be generally put into the following three categories.

1)Audio Segmentation and Classification. One basic problem in audio segmentation and classification is the discrimination between speech and music, because they are the two most important types of audio. The approach presented in [15] used only the average zero-crossing rate and the energy features, and applied a simple thresholding procedure. While in [16], thirteen features in time, frequency, and cepstrum domains, as well as more complicated classification methods (MAP, GMM, etc.) were used to achieve a robust performance. Both approaches reported real-time classification with accuracy rate over 90%. Since speech and music are quite different in spectral distribution and temporal change pattern, it is generally not very difficult to reach a relatively high level of discrimination accuracy.

A further classification of audio data may take other sounds, besides speech and music, into consideration. In [17], audio signals were classified into "music", "speech" and "others". Music was first detected based on the average length of time that peaks exist in a narrow frequency region, then speech was separated out by pitch tracking. This method was developed for the parsing of news stories. An acoustic segmentation approach was also proposed in [18], where audio recordings were segmented into speech, silence, laughter and non-speech sounds. They used cepstral coefficients as features and the hidden Markov model (HMM) as the classifier. The method was mainly applied to the segmentation of discussion recordings in meetings.

Research in [19] was aimed at the analysis of the amplitude, frequency and pitch of audio signals, as well as simulations of human audio perception so that results may be used to segment audio data streams and to recognize music. They also used these features to detect the sounds of shot, cry and explosion which might indicate violence.

2)Content-Based Audio Retrieval. One specific technique in content-based audio retrieval is query-by-humming, through which a song is retrieved by humming the tune of it. The approach in [20] defined the sequence of relative differences in the pitch to represent the melody contour and adopted the string matching method to search similar songs. It was reported that, with 10-12 pitch transitions, 90% of the 183 songs contained in an audio database could be discriminated.

A music and sound effect retrieval system was proposed in [21], where the Mel-frequency cepstral coefficients (MFCC) were taken as features, and a tree-structured classifier was built for retrieval. Since MFCC do not represent the timbre of sounds properly, this method in general failed to distinguish music and environmental sounds with different timbre characters. In the content-based retrieval (CBR) work of the Musclefish Company [22], they took statistical values (including means, variances, and autocorrelations) of several time- and frequency-domain measurements to represent perceptual features like loudness, brightness, bandwidth, and pitch. As merely statistical values were used, this method was only suitable for sounds with a single timbre.

A quick audio retrieval method using active search was presented in [23]. It was to search quickly through broadcast audio data to detect and locate sounds using reference templates, based on the active search algorithm and histogram modeling of zero-crossing features. The exact audio signal to be searched should be known *a priori* in this algorithm.

3)Audio Analysis for Video Indexing. In [24] and [25], audio analysis was applied to the distinction of five different video scenes: news report, weather report, basketball game, football game, and advertisement. The adopted features included the silence ratio, the speech ratio and the subband energy ratio, which were extracted from the volume distribution, the pitch contour, and the frequency domain, respectively. The multilayer neural network (MNN) and the hidden Markov model (HMM) were used as the classifier in the two papers, respectively. It was shown that, when using MNN, the method worked well in distinguishing among reports, games and advertisements, but had difficulty in classifying the two different types of reports and the two different kinds of games. While using HMM, the overall accuracy rate increased, but there were misclassifications among all the five sorts of scenes. They also applied the same set of audio features in distinguishing commercials from news reports in broadcast news [26].

In [27], audio characterization was performed on MPEG data (actually, the sub-band level data) for the purpose of video indexing. Audio was classified into dialog, non-dialog and silence intervals. Features were taken from the energy, pitch, spectrogram, and pause rate domains, and organized in a threshold procedure. There were somehow quite a few mistakes occurring in the classification between dialog and non-dialog intervals.

An approach for video indexing through music and speech detection was proposed in [28]. They applied image processing techniques to the spectrogram of audio signals. The spectral peaks of music were recognized by applying an edge-detection operator, and the speech harmonics were detected with a comb filter. They also presented two application systems to demonstrate the indexing method. One let users randomly access video while the other created condensations of dramas or movies by excerpting meaningful video segments based on the locations of music and speech.

Another field of research which is quite important for audio content analysis, and has provided, to some extent, theoretical and experimental support to some parts of this work is the *Audio Scene Analysis (ASA)*, which was named after the classic text of Bregman [29] who did pioneering work in this field. The goal of this filed is to understand the way the auditory system and the brain process complex sound environments, where multiple sources which change independently over time are present. Two subfields are dominant: auditory *grouping* theory, which attempts to explain how multiple simultaneous sounds are partitioned to form multiple "auditory images"; and auditory *streaming* theory, which attempts to explain how multiple sequential sounds are associated over time into individual cohering entities.

The approaches reported over the last 15 years in the ASA literature have been strongly functional and computational in nature. Brown and Cooke termed the discipline of constructing computer models to perform auditory source segregation as *Computational Audio Scene Analysis (CASA)* [30]. One example is the work by Weintraub who used a dynamic programming framework around Licklider's autocorrelation model to separate voices of two speakers whose voices interfere in a single recording [31]. His goal originated in speech recognition and enhancement. That is, he wanted to clean up speech signals to achieve a better speech recognition performance. Another example is the system built by Ellis, which aimed to analyze the sound and segregate perceptual components from noisy sound mixtures such as a "city-street ambience" [32]. The Structured Audio in MPEG-4 unifies many ideas and efforts in this field and provides semantic and symbolic descriptions of audio (the decoder is standardized while mature techniques for the encoder are still to be developed in the coming years) [33]. It will be useful for ultra low-bit-rate transmission, flexible synthesis, and perceptually based manipulation and retrieval of sounds.

2.3 WORK ON VISUAL CONTENT ANALYSIS

The first step in content analysis of an image sequence is to segment it into shots based on camera breaks. This technique is usually called *video scene segmentation.* In this system, we prefer to call it *shot change detection* so as to differentiate the two concepts of *scene* and *shot.*

A number of digital video segmentation methods have been developed based on the concept that only a fraction of picture elements change in the amplitude in consecutive frames. Various metrics have been proposed for video scene segmentation for both raw and compressed data. The metrics used to detect the difference between two frames can be classified broadly into four types: pixel or block comparison, histogram comparison (of grey levels or color code), comparison of DCT coefficients in MPEG-encoded video, and the subband feature comparison method [34]. Among these metrics, it is reported that methods based on histogram comparison are most effective because they are less sensitive to camera motions and object movements.

In general, abrupt camera transitions can be detected easily as the difference between two consecutive frames is rather large. However, problems arise when the transition is gradual (as produced by special camera effects such as wipe, dissolve, fade-in, fade-out) where the shot does not change abruptly but over a period of a few frames. Zhang *et al.* [6] proposed a histogram twin-comparison approach to detect both abrupt and gradual shot changes. This method requires two cutoff thresholds: one higher threshold for detecting abrupt transitions, and a lower one for gradual transitions. They also proposed a multi-pass procedure to improve the computational speed and accuracy. There are also attempts for short change detection in the compressed domain. For example, Meng *et al.* [35] considered a video shot that contains N frames. Each frame is represented by using an L-bin color histogram. They used the average histogram difference over N frames and the variance of histogram differences as the temporal features of video. A fast-changing shot tends to have a larger variance, while a slow-changing shot normally has a smaller variance.

3. SUMMARY OF THE PROPOSED SYSTEM

3.1 FRAMEWORK FOR VIDEO SEGMENTATION AND INDEXING

Framework of the proposed system for audiovisual data segmentation and annotation based on multimodal media content analysis is shown in Figure 1.2. The first step is to demultiplex the MPEG video bit stream into the parts of audio, image, and caption. Then, a semantic segmentation of the audio stream based on audio content analysis is conducted. We call such a segmented unit as "audio scene", and index it as pure speech, pure music, song, speech with the music background, environmental sound with the music background, silence, etc. based on the audio classification approach developed in this research. Next, the image sequence is segmented into shots based on the analysis of visual information, i.e. color histograms and motion vectors. A shot contains consecutive frames without camera breaks. Keyframes will be extracted from each shot, and color, shape and texture features of the keyframes will be analyzed to give visual index of each shot. Meanwhile, keywords will be detected from the closed caption in a video sequence to form the textual index.

An index table is built for each video sequence to contain the time interval of each segment, as well as the audio, visual and textual indexes. The structure of the index table is designed to be hierarchical and easy to access which represents the relationship of multi-modal feature descriptors. With this index structure, we are able to retrieve "the performance of a particular song by Michael Jackson" with the audio index of "song", the keyframe with Michael Jackson's image and keywords in the song. The automatically generated metadata are synchronized with the video sequence for transmission and storage. A filtering and retrieval mechanism is then built to select the video segments which match the user's interest according to the annotation, and to present these segments to the user.

3.2 CONTENT ANALYSIS OF THE AUDIO STREAM

A hierarchical system is developed in this work for content-based segmentation, classification and retrieval of the audio stream, which is illustrated in Figure 1.3. The system is composed of three stages, as described below.

Since the audio signal has a wide range of contents, including voice, music, and various kinds of sounds in the environment, effective classification must be done in a hierarchical way. We divide the generic audio data segmentation and classification task into two stages. In the first stage, audio signals are segmented and classified into the following basic types: speech, music, song, speech with music background, environmental sound with music background, six types of environmental sound, and silence. It is called the coarse-level audio segmentation and indexing. For this level, we use physical audio features including the energy function, the average zero-crossing rate, the fundamental frequency, and the spec-

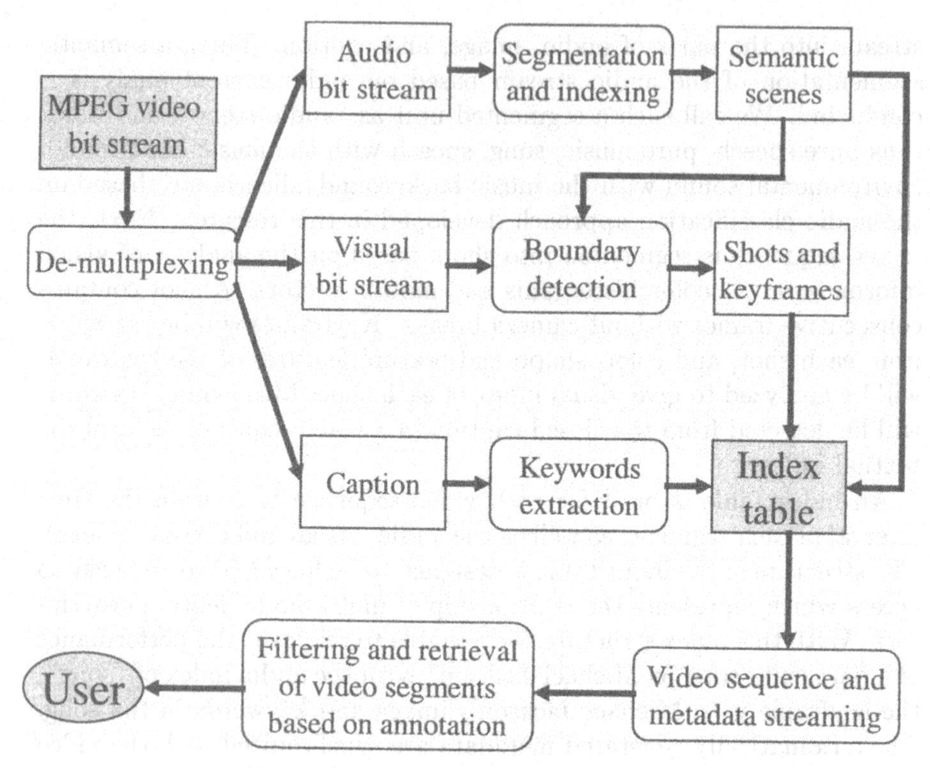

Figure 1.2. Framework for on-line segmentation and annotation of MPEG video.

tral peak tracks, which ensure the feasibility of real-time processing. We have worked on morphological and statistical analysis of temporal curves of these features to reveal differences among different types of audio. A rule-based heuristic procedure is built to classify audio signals based on these features.

Besides the commonly studied audio types such as speech and music in existing work, we have considered hybrid sounds which contain more than one basic audio type. For example, the speech signal with music background and the singing of a person are two types of hybrid sounds which have characters of both speech and music. We put these two kinds of sounds as separate categories in this stage, because they are very important in characterizing audiovisual segments. For example, in documentaries or commercials, there is usually a musical background with speech of commentary appearing from time to time. It is also common that clients want to retrieve the segment of video, in which there is singing of one particular song. There are other kinds of hybrid sounds such as speech or music with environmental sounds as the background

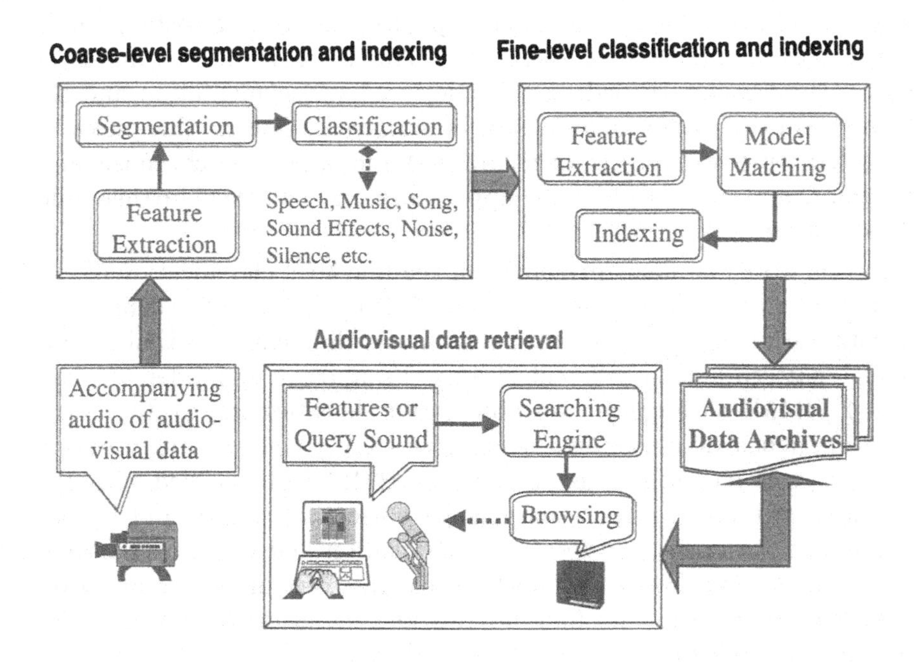

Figure 1.3. Hierarchical system for content-based segmentation, classification, and retrieval of the audio stream.

(where the environmental sounds may be treated as noise), or environmental sounds with music as the background.

In this stage, we also separate environmental sounds into six categories according to their harmony, periodicity, or stability properties. There are the "harmonic and fixed", "harmonic and stable", "periodic or quasi-periodic", "harmonic and non-harmonic mixed", "non-harmonic and stable", and "non-harmonic and irregular" environmental sounds. The approach for the first stage is model-free and can be applied, as the first step processing of audio data, to almost any content-based audio management system. Furthermore, it may be used as the tool for on-line segmentation and indexing of radio and TV programs. A table of content (TOC) may be generated automatically for each program, and the user is able to choose certain segments (e.g., those of pure music) to browse.

In the second stage, further classification is conducted within each basic type. For speech, we can differentiate it into the voices of man, woman, and child. For music, we classify it according to the instruments or types (for example, classics, blues, jazz, rock and roll, music with singing and the plain song). For environmental sounds, we classify them into semantic classes such as applause, bell ring, footstep, windstorm, laughter, birds' cry, and so on. This is known as the fine-level audio classification. Based on this stage, a finer segmentation and indexing result of audio material can be achieved. Due to differences in the origination of the three basic types of audio, i.e. speech, music and environmental sounds, different approaches can be taken in their fine classification.

In this work, we focus primarily on the fine-level classification of environmental audio. Features are extracted from the time-frequency representation of audio signals to reveal subtle differences of timbre and change pattern among different classes of sounds. The hidden Markov model (HMM) with continuous observation densities and explicit state duration densities is used as the classifier. Each kind of timbre in one audio class is represented as one state in HMM and modeled with the Gaussian mixture density. The change pattern of energy and timbre in the audio class is modeled by the transition and duration parameters of HMM. One HMM is built for each class of sound. The fine-level classification of audio data finds applications in automatic indexing and browsing of audio/video databases and libraries.

At the third stage of the system, an audio retrieval mechanism is built based on the archiving scheme described above. There are two retrieval approaches. One is query-by-example, where the input is an example sound, and the output is a ranking list of sounds in the database, which shows the similarity of the retrieved sounds to the input query. Similar to content-based image retrieval systems where image search can be done according to color, texture, or shape features, audio clips can also be retrieved with distinct features such as timbre, pitch, and rhythm. The user may choose one feature or a combination of features with respect to the sample audio clip. The other approach is query-by-keywords (or features), where various aspects of audio features are defined in a list of keywords. The keywords include both conceptual definitions (such as violin, applause, or cough) and perceptual descriptions (such as fastness, brightness, and pitch) of sounds. In an interactive retrieval process, the user may choose from a given menu a set of features, listen to the retrieved samples, and modify the input feature set accordingly to get a better matched result.

The hidden Markov model is used as the similarity measure in the query-by-example approach. That is, an HMM parameter set is trained for each sound clip in the database, and similar sounds are retrieved by matching the input query to each of the HMMs. As the database is organized according to the classification schemes, audio retrieval is more efficient. For example, the retrieval may be conducted only within certain classes. Also, with each retrieved sound, a class of audio data are found. Moreover, irrelevant or confusing results, as often appearing in image and video retrieval systems, can be avoided. Applications of audio retrieval may include searching sound effects in producing films, audio editing in making TV or radio programs, selecting and browsing materials in audiovisual libraries, and so on.

3.3 CONTENT ANALYSIS OF IMAGE SEQUENCES

In this research, we work on image sequences of the YUV format. A camera break detection scheme is proposed to segment an image sequence into shots based on histogram difference values between frames. It is developed based on the twin-comparison approach [6] with significant modifications on threshold selection and motion vector estimation. Especially, we incorporate histogram differences of the Y- and V- components to make the method much more robust. An adaptive procedure is also presented for the extraction of keyframes from each shot.

4. CONTRIBUTION OF THE RESEARCH

The following contributions have been made in the research of this monograph.

For audiovisual data segmentation and annotation based on multimedia content analysis:

- Instead of using a common model for video content, we observed structures and features of various video programs, and summarized rules for content modeling of different video types. A scheme is proposed to generate the annotation of audiovisual data based on individual video models. A brief classification of video data can also be conducted with these models, which will make the indexing and retrieval work more efficient and robust.

- Structures for two kinds of index tables are designed for audiovisual content. One is the primary index table which is hierarchically organized and chronologically ordered while the other is the secondary, non-linear and semantic index tree which allows fast browsing and retrieval.

For content-based audio segmentation, classification and retrieval:

- While existing research is normally focused on specific scenarios, a generic scheme is proposed in this work to cover all sorts of audio signals, including hybrid-type sounds and environmental sounds which are important in many real applications, but seldom considered in previous work.

- The proposed audio analysis system has a hierarchical and modular structure in which the physical and perceptual natures of different types of audio are well organized. It is flexible in the sense that each layer/module may be developed individually and has its own application domains.

- Feature extraction schemes are investigated based on the nature of audio signals and the problems to be solved. For example, the short-time features of energy, average zero-crossing rate and fundamental frequency are combined organically in distinguishing speech, music, environmental sound and silence. We use not only the feature values, but also their change patterns over the time and the relationship among the three kinds of features. We also propose the method for extracting spectral peak tracks, and use this feature specifically for the distinction of the sound segments of the song and the speech with a music background.

- Signal processing techniques are applied uniquely to the representation and classification of extracted features, including morphological and statistical analysis methods, the heuristic method, fuzzy logic, the clustering method, the hidden Markov model and the Gaussian mixture model, etc.

- Most audio features used in the system are short-time and one-dimensional, which provides the convenience for on-line audiovisual data processing.

- The coarse-level segmentation and classification approach is based on the observation of different types of audio signals and their physical features, which is generic and model-free. Thus, it can be easily applied to a wide range of applications.

- Although the coarse-level classification stage covers a wide range of audio signals, the complexity is low since the selected audio features are easy to compute and the rule-based indexing procedure is fast. Among the three short-time features, the fundamental frequency is the most expensive in computation, which only requires one 512-point FFT per 100 input samples. The spectral peak tracking needs a little bit more calculation, but it only has to be computed under certain conditions.

- In the fine-level classification of the environmental sound, the hidden Markov model is used to build a bridge over the gap between low-level features and semantic meanings of the sound. We also investigate mathematical models for representing the perceptual features of timbre and rhythm of environmental sound.
- The content-based audio retrieval scheme is built on top of the audio classification results, thus leading to semantic meanings and better reliability. It is also highly efficient in association with well organized archives.

For content analysis of image sequences:

- A highly efficient and robust method for shot change detection is developed based on histogram comparison. Significant modifications have been made to the existing approach. The new scheme is designed for image sequences of the YUV format, and the compensative relation between Y- and V- components are exploited. An adaptive algorithm for keyframe extraction is also proposed and proved to be very effective.

In general, compared to existing work in related areas, the proposed research is among the most advanced in both the scope and quality for solving the problem.

5. OUTLINE OF THE MONOGRAPH

This monograph is organized as follows. In Chapter 2, data structures and regularities of different video types are analyzed. A scheme for generating annotation of the audio and visual content is proposed. In Chapter 3, computations and characteristics of audio features used in this research are introduced. The proposed procedures for the coarse-level segmentation and indexing of generic audio data are presented in Chapter 4. Concepts, algorithms and implementational issues for fine-level classification and retrieval of environmental sounds are described in Chapter 5. Studies on visual shot change detection and keyframe extraction are presented in Chapter 6. Experimental results of proposed algorithms are presented in Chapter 7. Finally, concluding remarks and future research plans are given in Chapter 8.

II
VIDEO CONTENT MODELING

Chapter 2

VIDEO CONTENT MODELING

1. COMMON MODEL FOR VIDEO CONTENT

The common model used for video data is a hierarchical structure consisting of scenes, shots, and frames [34]. A shot is a pictorial unit defined as a set of contiguously recorded image frames. It is associated solely with the visual information, and each shot has a clear physical scope. A scene is rather a semantic concept which refers to a relatively complete paragraph of video having coherent semantic meanings. It is composed of one or more consecutive shots. A scene may have its visual, audio, and textual content, and is normally more subjectively defined. However, there is a need to give a consistent definition of the scene for modeling the video content.

Different video types may have different associated semantics. Thus, a video content model should be built according to features of each video type individually. For example, newscast video has very different structure from those of movies. It would be beneficial that we understand the semantics present in the data type for data modeling. This facilitates efficient and effective data retrieval. Moreover, with suitable data models, video content can be classified into types such as newscast, sports video, conference, etc., which are important in applications like user agent driven media selection and filtering, video-on-demand and semantic multicast backbone video. With these models, we also know which features to be extracted from which types of data, thereby raising the processing efficiency and reducing the information to be stored.

In this work, we observed video data belonging to the following five types: news bulletins, documentaries (such as scientific videos and educational videos), feature movies (including TV drama series), variety shows, and sports video. The characteristics of each video type are summarized below.

2. MODELS FOR DIFFERENT VIDEO TYPES

2.1 NEWS BULLETIN

A data model for news video was examined in [36]. The news bulletin is a simple sequence of news items that are possibly interleaved with commercials. A news item in most news programs always starts with an anchorperson shot, followed by a sequence of shots which illustrate the news story. Frames of the anchorperson shots have a well defined spatial structure. Thus, the scene concept in news video is relatively simple. It is composed of news items, and each news item has one or more shots. We also observe that, in terms of both audio and visual properties, each news item normally starts with not only the image but also the speech of the anchorperson. Then, there may be shots of other sights with the anchorperson's voice as the offscene sound.

In one such example, there are eight shots in the news item. One keyframe is extracted from each shot as shown in Figure 2.1. The first shot is the anchorperson shot, and the voice of the anchorperson runs through the whole item. Sometimes there are also voices of on-site reporters, speeches in interviews, as well as environmental sounds in one news item. However, the picture and the voice of the anchorperson will be back when the next news item begins. Therefore, by recognizing the voice and the image of the anchorperson, one can detect breaks between news items easily. The voice of an anchorperson is usually fast, stable and clear. Sometimes, the video shot may not change over two consecutive news items and simply stays at the anchorperson. However, there is obvious pause in the anchorperson's speech between two news items, by which one still can detect breaks of news items. After news item segmentation, each item can be represented by keyframes of visual shots as well as the first one or two sentences of the anchorperson. Commercials in the middle of one news program can be detected and separated out by its audio features, i.e. there is normally music or special sound effects with scattered human speeches now and then. The weather report at the end of the news program may be characterized by a keyframe in which the weather reporter speaks with a map in the background.

2.2 VARIETY SHOW VIDEO

Similar to news bulletin, the variety show video does not have complicated scenes either. It is mainly composed of a sequence of performances.

Figure 2.1. Keyframes extracted from shots within one TV news item.

There are normally music and/or songs during one performance. A performance usually begins with some pure music. At the end of each performance, there are the pause of music, the applause and acclaim from the audience, and the speech of the host. Sometimes, the music does not stop between two performances. However, there is a change of the melody and the rhythm of the music which can be detected. By looking for these audio clues, we can segment the video program into individual performances and transitional parts (e.g. interviews with the actor or the audience by the host). Each performance can be indexed by keyframes of the actor(s) as well as the audio semantics (such as pure music, song, etc.). An example is shown in Figure 2.2, where keyframes and the associated audio index of one performance in a variety show video are displayed.

2.3 SPORTS VIDEO

In ball games (such as basketball, soccer, volleyball) of sports video, sounds of whistle may represent the start and/or end of one episode. Bursts of applause and acclaim from the audience may indicate some exciting moments. Also, the color of the floor (normally yellow in basketball and volleyball games) or the grass may indicate shots of the game (rather than shots of the audience, of a particular player or of some other sights, as shown in Figure 2.3). This dominant color feature can be easily detected from the color histograms. Thus, combining the audio and

Figure 2.2. Keyframes and the associated audio index of one performance in a variety show video.

visual information, we can detect breaks of episodes as well as exciting moments within one game.

In golf games, the player normally stands quite still adjusting the position of club for a while before hitting the ball (which is reflected as very little changes in the histograms of neighboring frames), and then suddenly hit the ball (reflected as bigger changes in the neighboring histograms). After one good hit, there are applause and acclaim from the audience. Therefore, with these *a priori* knowledges, we can find the segments of wonderful hits easily.

Figure 2.3. Keyframes of shots for soccer and basketball games from sports video.

2.4 DOCUMENTARIES

In documentary movies and videos, there are the structures of semantic scenes which are difficult to define simply by using audio and visual features. However, with the help of audio clues, segmentation results can be much more improved than using the visual information alone. In such kind of video, audio parts are accompanied with pictorial parts off-line. Normally, there is music all through the program with commentary speech appearing from time to time. There are often obvious pauses or changes of audio contents at scene breaks. For example, the long pause between two speech segments, the variation or stop of music may all indicate change of scenes.

Let us give an example below. There is a 33-second long clip in a documentary movie which provides a scene of the life style in Lancaster. There are seventeen shots in the scene displaying various aspects of people's life in Lancaster. However, the accompanying music episode, which is continuous during the period and isolated from the former and the latter episodes by obvious pauses, shows that these shots are within one semantic unit. Some keyframes of shots in this scene are shown in Figure 2.4.

Figure 2.4. Keyframes within one scene ("Lifestyle in Lancaster") of diverse shots in a documentary movie.

2.5 FEATURE MOVIES AND TV SERIES

In feature movies and TV drama series, the situation is even more complicated because their content structures are of a greater diversity. Nevertheless, there are many scenes of conversations in such video types which can be detected by tracing voices and shots of speakers. An example is shown in Figure 2.5 where the keyframes of several consecutive shots within one scene of conversation in a TV drama series are displayed. We can see that there are shots of the two speakers alternately. Sometimes, there is also music as the offscene sound with multiple shots appearing, and the scene can be defined with such an audio feature. In certain TV drama series, there is a fixed short music clip at each scene break. Most of the time, there are several shots within one audio scene (defined as a unit of video with continuous audio content), but there are also situations that the audio content may change (e.g. the start or stop of music, the pause of speech) within one visual shot. Therefore, a scene break can only be defined if both audio scene change and visual shot change occur.

Figure 2.5. Keyframes from different shots within one scene of conversation in a TV drama series.

3. PROPOSED SCHEME FOR VIDEO CONTENT PARSING

Based on observations described above, we propose a scheme for video content parsing by using video models together with audio and visual information analysis as illustrated in Figure 2.6. The video input is first demultiplexed into the audio data stream and the image sequence. Then, audio data are segmented into audio scenes, and each scene is classified as one of the basic audio types. The image sequence is broken into shots, and keyframes are extracted from each shot. Meanwhile, facts about the video data including the video type and other *a priori* knowledges, which are either metadata accompanying with the video data or the input from the user, are used to select a data structure from the video model database.

The video model database consists of data structures for various video types. The data structure provides a guidance for video classification, index table generation, as well as special audio and visual feature analysis associated with a particular video model. For example, the detection of sounds of whistle, applause and acclaim, and the search for shots of green grass are important for parsing video clips of soccer and football games. Included in the data structure are the syntax of the index table, and models for characteristic sounds, shots, and keyframes of the specific video type. Finally, The audio scene indices, the keyframes of each shot, as well as the special audio and visual analysis results are integrated to build the index table according to the video model selected for this audiovisual sequence.

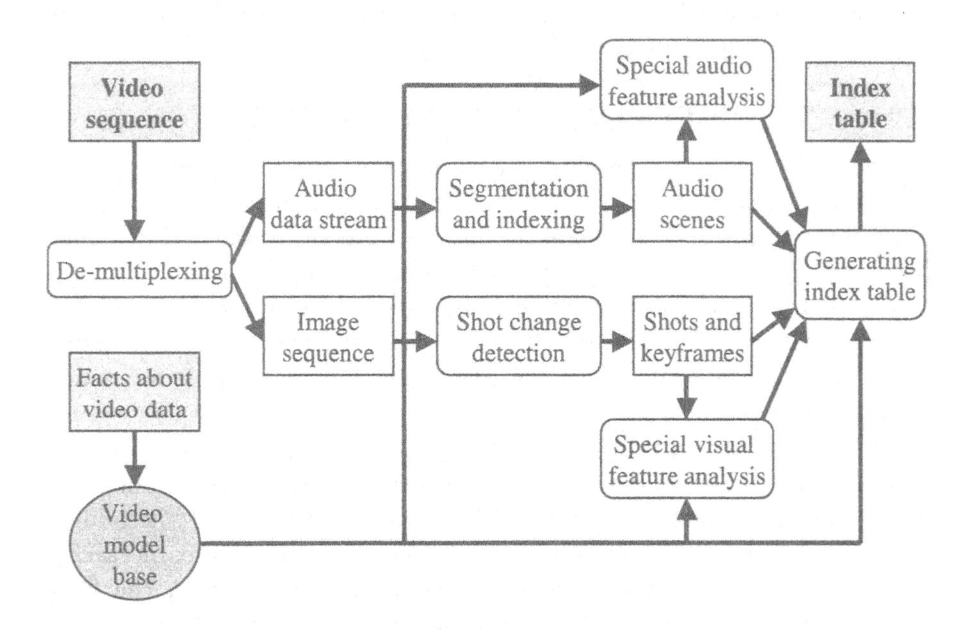

Figure 2.6. The framework for video content parsing with combined audio and visual information analysis.

4. DESIGN OF INDEX TABLE FOR NON-LINEAR ACCESS

4.1 THE PRIMARY INDEX TABLE

The primary index table of a video package is produced automatically at run time, and is chronologically ordered. In general, it has a hierarchical structure consisting of three levels: the package-level, the scene-level and the shot-level. Included in the first level are metadata about the whole video package such as the duration, the format and the model type of the video. At the second level, there is information about each scene including the time interval and the semantic audio and visual indices. Then, for each shot, the time interval and some low-level indices are organized in the third level.

While this kind of index table is easy to generate on-line, it provides only linear access to video data that may not be optimal for fast browsing and retrieval. For example, in order to retrieve a particular segment of male solo performance, we have to search from the beginning of the video package. Thus, besides the primary index table, there is the need to design a second kind of content index form for this system, which provides non-linear access to the video source.

4.2 THE SECONDARY INDEX TREE

One approach based on clustering was proposed in [37] where video shots were clustered according to color histograms of keyframes and some other features. The clustering was done hierarchically and included fuzzy logic (i.e. one shot might be put into two or more classes). Finally, a tree structure of video shots was obtained which was convenient for video browsing and retrieval. However, this structure was constructed based on only low-level features which might not be suitable for many application scenarios.

We propose here a preliminary design of a secondary index form for video content which is derived from the primary index table. For each video clip, this index form will include three index trees for the audio, visual, and textual information, respectively. The audio index hierarchy will be built according to semantic meanings. For example, the first level of the tree may include nodes indexed as "pure speech", "pure music", "song", "environmental sound", "silence", etc. And the node of "pure speech" may have a second level hierarchy including "male speech", "female speech", "conversation" and so on. Under each node, there are time intervals of segments within the video sequence which belong to the audio type indexed by this node. An illustration of the audio index tree is shown in Figure 2.7.

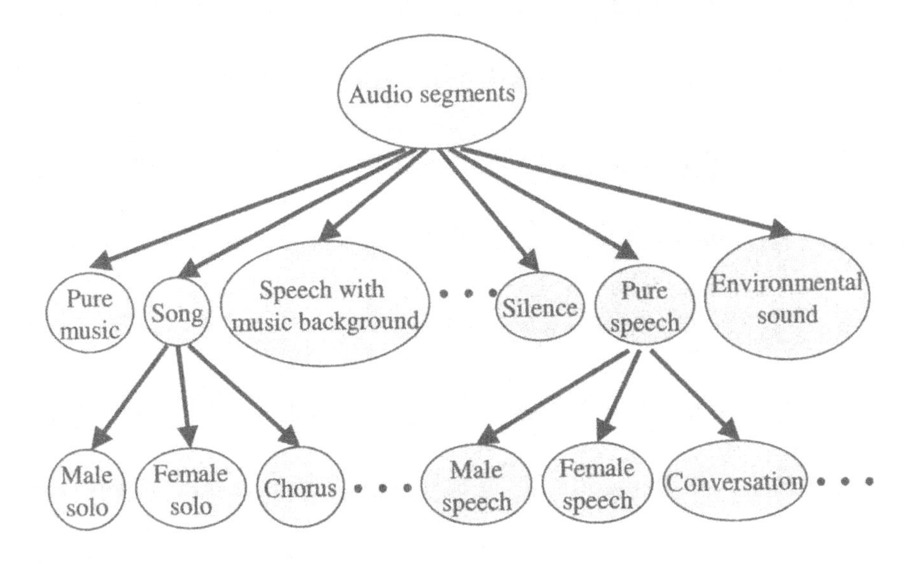

Figure 2.7. **Illustration of an audio index tree as part of the secondary, non-linear index structure.**

The textual index tree may be organized alphabetically or according to a subject list containing keywords. Each node in the tree is indexed with a keyword (e.g. "economy", "football", "Bill Clinton"), and maintains time intervals of segments in the video source which may be indexed with this keyword. For visual information, it is quite hard to find some semantic way to index it. One possibility is to analyze image features of keyframes from each shot, and classify them into categories such as "background image", "texture image", and "object image". Then at the second level, the "object image" may be further separated into clusters including "people", "animal", "building" and so on. And the "background image" may include "scenery", "crowd", etc. In this way, the visual index tree can be built (as illustrated in Figure 2.8). The user may choose to browse or retrieve video segments according to one type of index or a combination of several types of indexes.

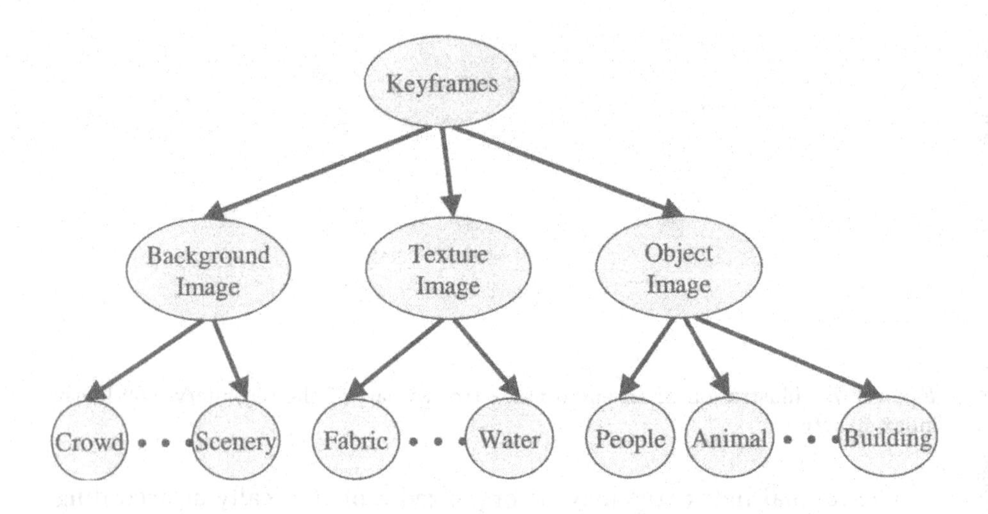

Figure 2.8. Illustration of a visual index tree as part of the secondary, non-linear index structure.

III
AUDIO CONTENT ANALYSIS

Chapter 3

AUDIO FEATURE ANALYSIS

There are, in general, two types of audio features: the physical features and the perceptual features. Physical features refer to mathematical measurements computed directly from the sound wave, such as the energy function, the spectrum, the cepstral coefficients, the fundamental frequency, and so on. Perceptual features are subjective terms which are related to the perception of sounds by human beings, including loudness, brightness, pitch, timbre, rhythm, etc. In this research, we use the temporal curves of three kinds of short-time physical features, i.e. the energy function, the average zero-crossing rate, and the fundamental frequency, as well as the spectral peak tracks for the purpose of coarse-level audio segmentation and classification. While for the fine-level classification and retrieval, one of our most important tasks is to build physical and mathematical models for the perceptual audio features with which human beings distinguish the subtle differences among different classes of sounds. Currently, we consider two kinds of perceptual features in this work, i.e. timbre and rhythm.

1. AUDIO FEATURES FOR COARSE-LEVEL SEGMENTATION AND INDEXING OF GENERIC DATA

1.1 SHORT-TIME ENERGY FUNCTION

The short-time energy function of an audio signal is defined as

$$E_n = \frac{1}{N} \sum_m [x(m)w(n-m)]^2, \tag{3.1}$$

where $x(m)$ is the discrete time audio signal, n is time index of the short-time energy, and $w(m)$ is a rectangle window, i.e.

$$w(n) = \begin{cases} 1, & 0 \leq n \leq N-1, \\ 0, & \text{otherwise.} \end{cases}$$

It provides a convenient representation of the amplitude variation over the time. By assuming that the audio signal changes relatively slowly within a small interval, we calculate E_n once every 100 samples at an input sampling rate of 11025 samples per second (i.e. compute E_n at around every 10ms). We set the window duration of $w(n)$ to be 150 samples so that there is an overlap between neighboring frames. The audio waveform of a typical speech segment and the temporal curve of its short-time energy function are shown in Figure 3.1. Note that the sample index of the energy curve is at the ratio of 1:100 compared to the corresponding time index of the audio signal.

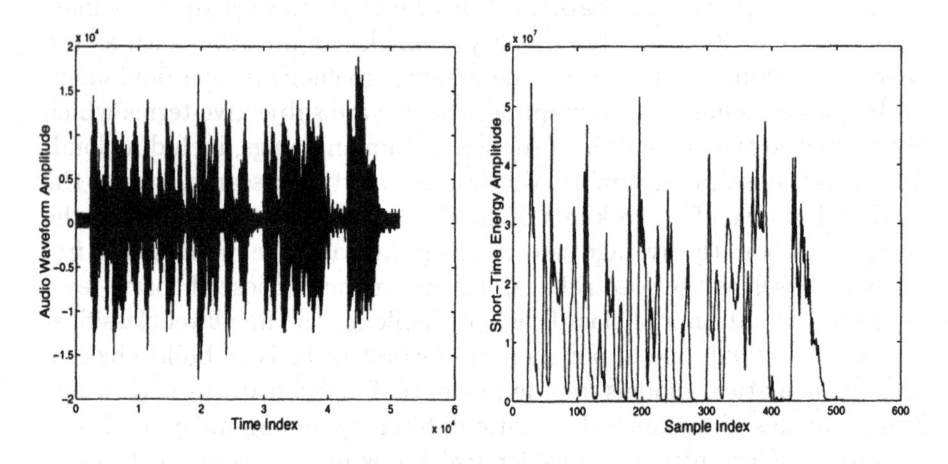

Figure 3.1. The audio waveform and the short-time energy function of a speech segment.

The major significances of using the short-time energy feature in our work include: (1) For speech signals, it provides a basis for distinguishing voiced speech components from unvoiced speech components. This is due to the fact that values of E_n for the unvoiced components are in general significantly smaller than those of the voiced components, as can be seen from the peaks and troughs in the energy curve. (2) It can be used as the measurement to distinguish audible sounds from silence when the signal-to-noise ratio is high. (3) Its change pattern over the time may reveal the rhythm and periodicity properties of sound.

1.2 SHORT-TIME AVERAGE ZERO CROSSING RATE

In the context of discrete-time signals, a zero-crossing is said to occur if successive samples have different signs. The rate at which zero-crossings occur is a simple measure of the frequency content of a signal. This is particularly true of narrowband signals. Since audio signals may include both narrowband and broadband components, the interpretation of the average zero-crossing rate is less precise. However, rough estimates of spectral properties can still be obtained using a representation based on the short-time average zero-crossing rate (ZCR), as defined below:

$$Z_n = \frac{1}{2}\sum_m |sgn[x(m)] - sgn[x(m-1)]|w(n-m), \tag{3.2}$$

where

$$sgn[x(n)] = \begin{cases} 1, & x(n) \geq 0, \\ -1, & x(n) < 0, \end{cases}$$

and

$$w(n) = \begin{cases} 1, & 0 \leq n \leq N-1, \\ 0, & \text{otherwise}. \end{cases}$$

Temporal curves of the short-time average zero-crossing rate for several audio samples are shown in Figure 3.2. Similar to the computation of the short-time energy function, we also choose to compute the ZCR at every 100 input samples, and set the window width to 150 samples.

The speech production model suggests that the energy of voiced speech signals is concentrated below 3 kHz because of the spectral fall-off introduced by the glottal wave, whereas most of the energy is found at higher frequencies for unvoiced speech signals [38]. Since high (or low) frequencies imply high (or low) zero-crossing rates, a reasonable rule is that if the zero-crossing rate is high, the speech signal is unvoiced while if the zero-crossing rate is low, the speech signal is voiced. Hence, the zero-crossing rate can be used for making distinction between voiced and unvoiced speech signals. As shown in Figure 3.2(a), the speech ZCR curve has peaks and troughs from unvoiced and voiced components, respectively. This results in a large variance and a wide range of amplitudes for the ZCR curve. Note also that the ZCR waveform has a relatively low and stable baseline with high peaks above it.

Compared to that of speech signals, the ZCR curve of music plotted in Figure 3.2(b) has a much lower variance and average amplitude, suggesting that the average zero-crossing rate of music is normally much more stable during a certain period of time. ZCR curves of music generally have an irregular waveform with a changing baseline and a relatively small range of the amplitude.

Since environmental audio consists of sounds of various origins, their ZCR curves can have very different properties. For example, the average zero-crossing rate of the sound of chime as shown in Figure 3.2(c) reveals a continuous drop of the frequency centroid over the time while that of the footstep sound in Figure 3.2(d) is rather irregular. We may briefly classify environmental sounds according to the properties of their ZCR curves such as regularity, periodicity, stability, and the range of amplitudes.

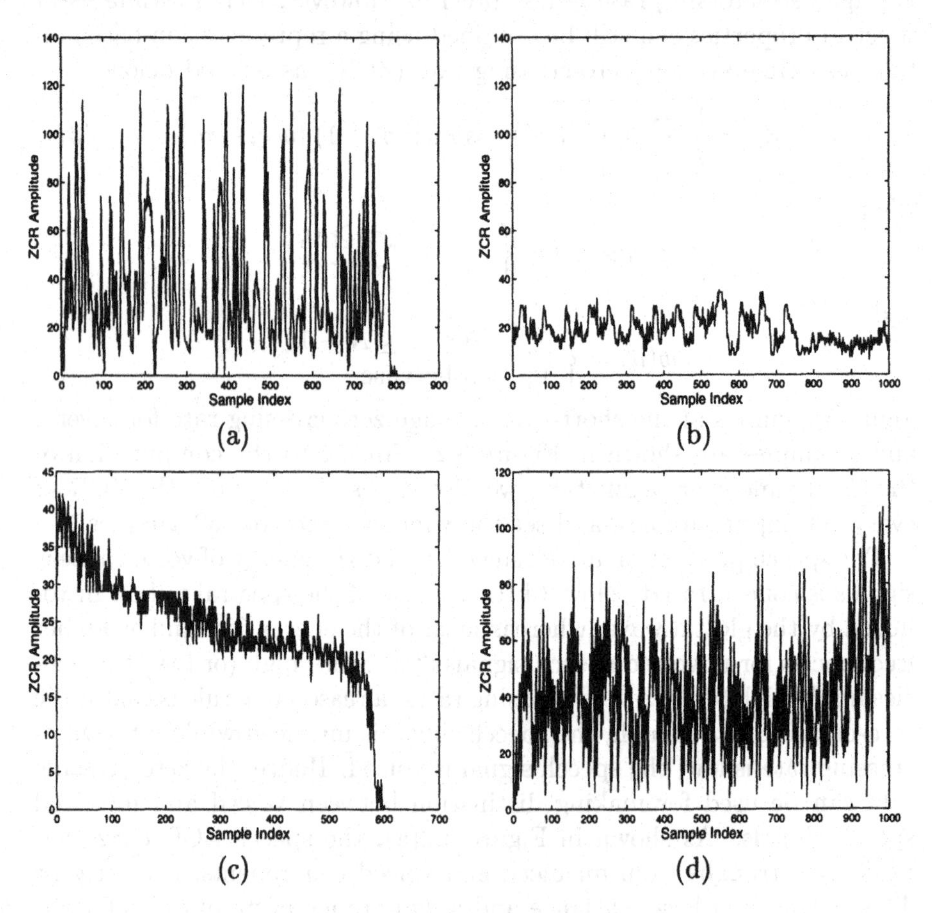

Figure 3.2. The short-time average zero-crossing rates of four audio signals: (a) speech, (b) piano, (c) chime and (d) footstep.

1.3 SHORT-TIME FUNDAMENTAL FREQUENCY

A harmonic sound consists of a series of major frequency components including the fundamental frequency and those which are integer multiples of the fundamental one. With this concept, we may divide sounds into two categories, i.e. harmonic and non-harmonic sounds. The spectra of sounds generated by violin and applause are illustrated in Figure 3.3, respectively. It is clear that the former one is harmonic while the latter one is non-harmonic.

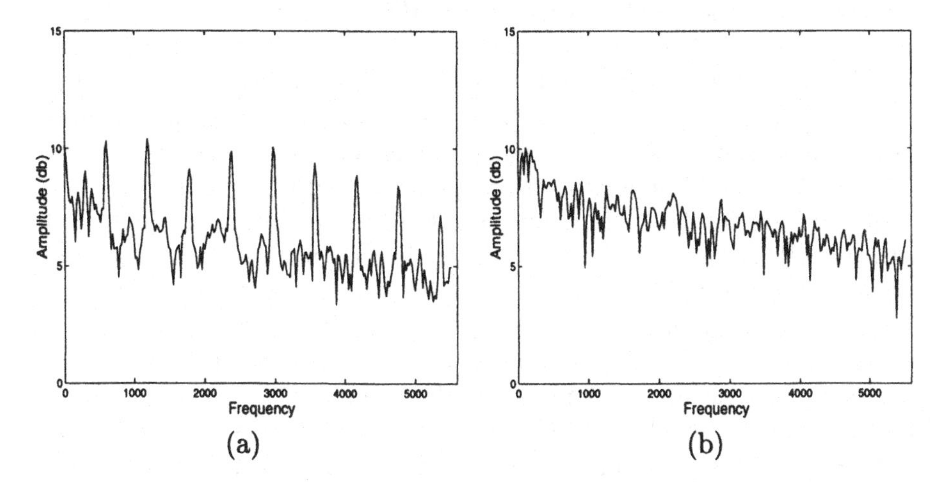

Figure 3.3. Spectra of harmonic and non-harmonic sound computed directly with FFT: (a) violin and (b) applause.

Whether an audio segment is harmonic or not depends on its source. Sounds from most musical instruments are harmonic. The speech signal is a harmonic and non-harmonic mixed sound, since the voiced components are harmonic while the unvoiced components are non-harmonic. Most environmental sounds are non-harmonic, such as the sounds of applause, footstep, and explosion. But there are also examples of sound effects which are harmonic and stable, like the sounds of doorbell and touch-tone; and those which are harmonic and non-harmonic mixed like laughter and dog bark.

In order to measure the harmony feature of sound, we define the short-time fundamental frequency (SFuF) as such: when the sound is harmonic, the SFuF value is equal to the fundamental frequency estimated from the audio signal; and when the sound is non-harmonic, the SFuF is set to zero.

Fundamental frequency estimation, or equivalently pitch detection, has been one of the most important problems in both speech signal processing [38] and music content analysis [39] - [41] (It is however worthwhile to point out that the fundamental frequency is a physical measurement while the pitch is rather a perceptual term which is analogous to the frequency but not exactly the same, as stated in [42]). There are many schemes proposed to solve this problem, but none of them is perfectly satisfactory for a wide range of audio signals. Our primary purpose of estimating the fundamental frequency is to detect the harmonic property for all kinds of audio signals. Thus, we tend to build a method which is efficient, robust, but not necessarily perfectly precise.

In this work, the short-time fundamental frequency is calculated based on peak detection from the spectrum of the sound. The spectrum is generated with autoregressive (AR) model coefficients estimated from the autocorrelation of audio signals. This AR model generated spectrum is a smoothed version of the frequency representation. Moreover, as the AR model is an all-pole expression, peaks are prominent in the spectrum. Comparing the spectra shown in Figure 3.4 which were generated with the AR model with those computed directly from the FFT of audio signals as shown in Figure 3.3, we can see that detecting peaks associated with the harmonic frequencies is much easier in the AR generated spectrum than in the directly computed one. In order to keep a good precision of the estimated fundamental frequency, we choose the order of the AR model to be 40. With this order, harmonic peaks are remarkable while there are non-harmonic peaks appearing. However, compared with harmonic peaks, non-harmonic ones not only lack a precise harmonic relation among them, but also are usually less sharp at the maximum and are of smaller heights. One example of such comparison is clearly shown in Figure 3.4. Therefore, for a sound to be regarded as harmonic, there should be the least-common-multiple relations among the peaks, and at least some of the peaks should be sharp and high enough.

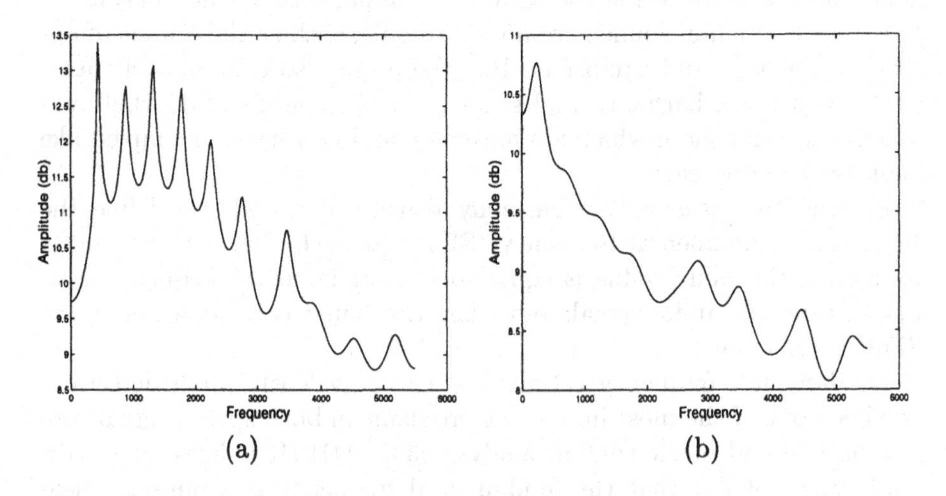

Figure 3.4. Spectra of harmonic and non-harmonic sound generated with AR model: (a) trumpet and (b) rain.

The computation of the spectrum generated by the AR model is as follows. First, the short-time autocorrelation of an audio signal can be computed as

$$R_n(k) = \frac{1}{N}\sum_{m}[x(m)w(n-m)][x(m+k)w(n-m+k)], \quad 0 \leq k \leq P, \tag{3.3}$$

where $w(m)$ is the rectangle window of width N, and P is the order of the AR model. Here, we choose the window width to be 512 samples. Second, the AR model parameters (including the prediction coefficients $a_1, a_2, \ldots, a_P$, and the mean-square prediction error E_P) are estimated from the values of $R_n(k)$ through the Levinson-Durbin algorithm [43]. Then, the power spectrum of the signal can be estimated as:

$$\hat{S}_{AR}(e^{j\frac{2\pi}{N}l}) = \frac{E_P}{|1+\sum_{k=1}^{P} a_k e^{-j\frac{2\pi}{N}kl}|^2} = \frac{E_P}{|\sum_{k=0}^{N-1} a_k e^{-j\frac{2\pi}{N}kl}|^2},$$

$$0 \leq l \leq N-1, \tag{3.4}$$

where $a_0 = 1$ and $a_{P+1}, a_{P+2}, \ldots, a_{N-1}$ are all set to zero. Thus, the denominator of $\hat{S}_{AR}$ can be computed with an N-point FFT. We choose N to be 512, because almost all sound harmonics can be revealed under such a frequency resolution. Finally, the logarithm of the square-root of each $\hat{S}_{AR}$ value is calculated.

All maxima in the spectrum are detected as potential harmonic peaks, and the amplitude, the width, and the sharpness of each peak are calculated. The sharpness of a peak is computed as the second order difference at the maximum. The amplitude at each side (left or right) of the peak is the difference between the spectrum value at the maximum and the spectrum value at the inflexion at that side. The width is the distance between the two inflexions. The inflexion at one side of the peak is defined to be the first point from the maximum at that side that has a slope less than 50% of the slope of its previous point. It is checked among locations of these peaks whether a certain amount of them have a common divider and at least some of them have sharpness, amplitude, and width values satisfying certain criteria. If all conditions are met, the SFuF value is estimated as the frequency corresponding to the common divider of locations of harmonic peaks. Otherwise, the SFuF is set to zero. SFuF is computed once every 100 input samples. After the temporal curve of SFuF is obtained for a segment of a certain length, there is a post-processing step in which singular points in the temporal curve of SFuF are removed to improve the accuracy of the SFuF estimation.

Illustrated in Figure 3.5 are examples of the SFuF curves of sounds. Shown on top of each picture is the "zero ratio" of the SFuF curve for that sound segment, which is defined as the ratio between the number of samples with a zero SFuF value (i.e. the non-harmonic sound) and the total number of samples in the curve. We see that music is generally continuously harmonic. Also, the fundamental frequency usually changes more slowly than that of other kinds of sounds, and the SFuF value tends to concentrate on certain values for a short period of time. Harmonic and non-harmonic components appear alternately in the SFuF curve of the speech signal, since the voiced components are harmonic and the unvoiced components are non-harmonic. The fundamental frequency of the voiced components is normally in the range of 100-300Hz. Most environmental sounds are non-harmonic with zero ratios over 0.9. The sound of rain is one example of them. An instance of the harmonic and non-harmonic mixed sound effects is the sound of laughing, in which the voiced segments are harmonic, while the intermissions in between as well as the transitional parts are non-harmonic. It has a zero ratio of 0.25 which is similar to that of the speech segment.

1.4 SPECTRAL PEAK TRACK

The peak tracks in the spectrogram of an audio signal may often reveal some characteristics of the type of sound. For example, sounds from musical instruments normally have spectral peak tracks which remain at the same frequency level and last for a certain period of time as shown in Figure 3.6. The spectrogram images in this figure were originally represented in form of pseudo color with the x-axis denoting time in second and the y-axis denoting frequency in Hz. The pixel color denotes the amplitude of FFT coefficients. The warmer the pixel color looks, the higher amplitude is in the frequency domain. Sounds from human voices have harmonic peak tracks in their spectrograms which align tidily in the shape of a comb. The spectral peak tracks in songs may exist in a broad range of frequency bands, and the fundamental frequency ranges from 87Hz to 784Hz (including voices of bass, tenor, alto, and soprano). There are relatively long tracks in songs which are stable because the voice stays at a certain note for a period of time, and they are often in a ripple-like shape due to the vibration of vocal chords. The spectral peak tracks in speech normally lie in the lower frequency bands, and are more close to each other due to the fundamental frequency range of 100-300Hz. They also tend to be of a shorter length because there are intermissions between voiced syllables, and they may fluctuate slowly because the pitch may change during the pronunciation of certain syllables.

We extract spectral peak tracks for the purpose of characterizing sounds of song and speech. Basically, it is done by detecting peaks in the power spectrum generated by the AR model parameters and checking

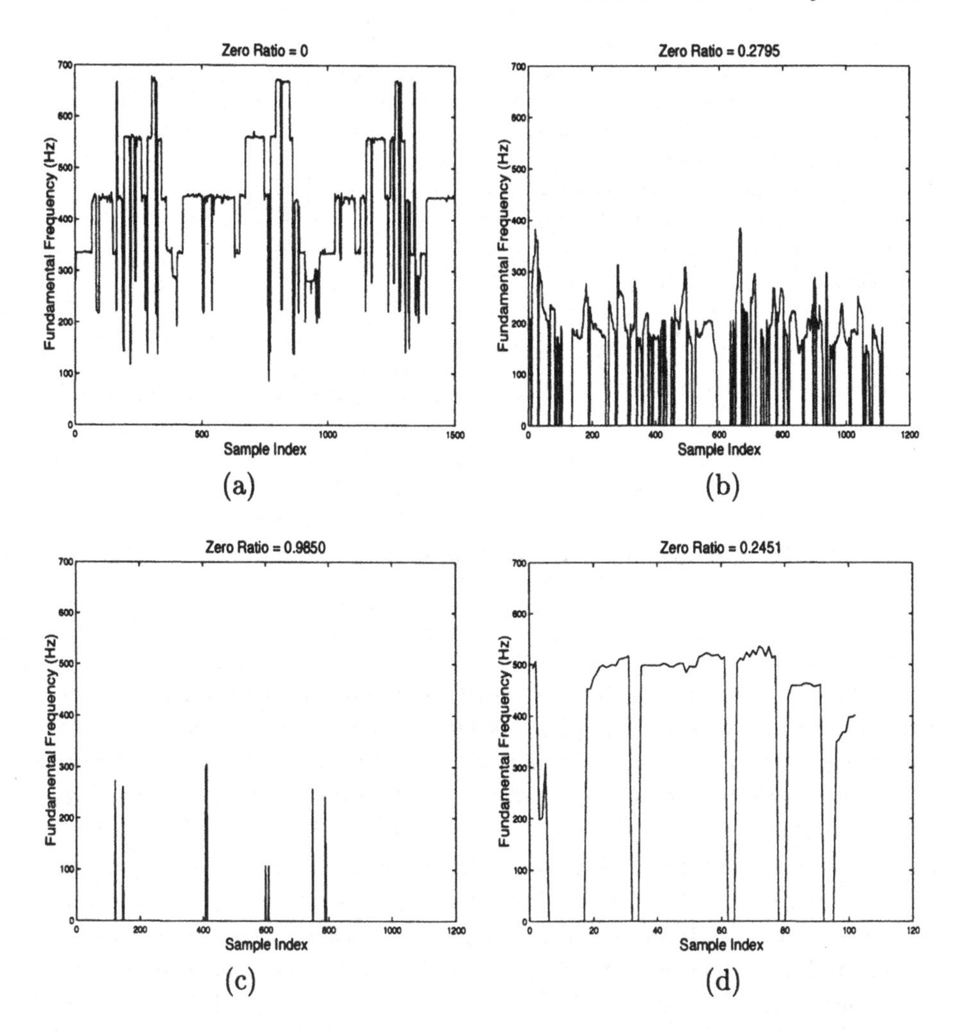

Figure 3.5. The short-time fundamental frequency of audio signals: (a) trumpet, (b) speech, (c) rain and (d) laugh.

harmonic relations among the peaks. Compared to the problem of fundamental frequency estimation where the precision requirement is less strict and slight errors are allowed, the task here is more difficult in the sense that locations of tracks should be determined accurately. It should be the best that all tracks are detected without any artifact, which is very difficult to achieve. Nevertheless, with the confinement that only spectral peak tracks in song and speech segments are considered and based on distinct features of such tracks as mentioned above, we derived a set of rules to pick up proper harmonic peaks.

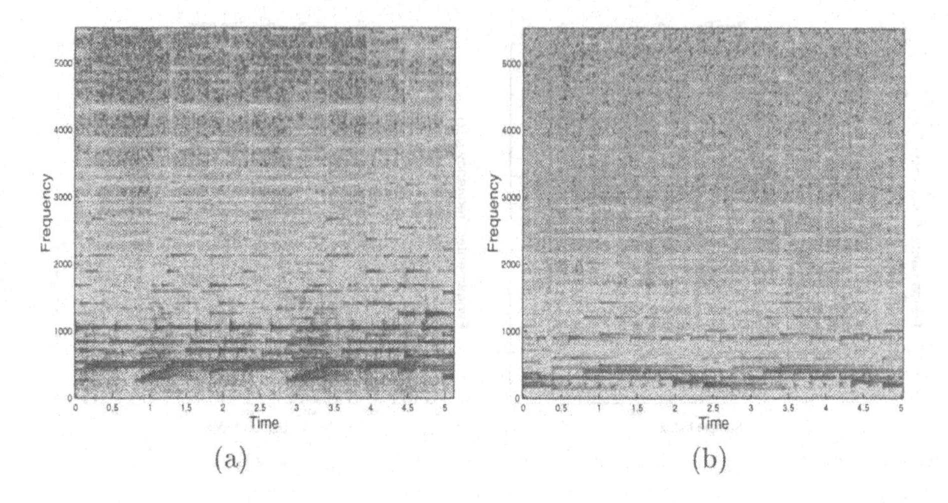

Figure 3.6. The spectrograms of sounds from musical instruments: (a) piano and (b) guitar.

We set the range of the fundamental frequency of harmonic peaks under consideration to 80Hz-800Hz. With a 512-point FFT, the frequency resolution should be enough for such a range if the order of the AR model is chosen properly. The critical issue is to choose the order P of the AR model so that harmonic peaks are revealed in the generated spectrum and easy to be detected. For example, when $P = 40$, harmonic peaks with a fundamental frequency higher than 250Hz can be easily detected. This condition fits for most song segments. However, this resolution is not enough for spectral peaks in most male and female speech segments. By experiments, we found that $P = 80$ was normally suitable for female speech signals (with a fundamental frequency at about 150-250Hz), and male speech signals might require an order of $P = 100$ when the fundamental frequency is between 100-150Hz. However, with these higher values of P, many artifact peaks will appear in the estimated spectrum of sounds having higher fundamental frequencies, and may severely impair the quality of peak detection in these sounds.

We used the final prediction error criterion and Akaika's information criterion to automatically estimate the optimal order of the AR model for a frame of an audio signal. Unfortunately, these criteria did not always work properly for this purpose. In order to get the best detection precision possible, we currently fix the order of the AR model at three levels, i.e. 40, 80, and 100. The main idea is that it should be able to detect harmonic peaks with one of these orders for sounds of concern. The procedure to determine the proper order is stated below. If, in the previous frame of an audio signal, harmonic peaks were detected from the power spectrum generated with the AR model of order P_1 (P_1 may be 40, 80, or 100), we begin to detect harmonic peaks for the current signal frame with the spectrum of order P_1. If harmonic peaks are found in this spectrum, then we go on to the next signal frame. Otherwise, we will try the spectrum generated with the other two order levels. If no harmonic peaks were detected in the previous frame, we try the three order levels one by one for the current frame until harmonic peaks are found or the conclusion of no harmonic peaks existing is obtained. Harmonic peaks should have harmonic relations among them and satisfy some sharpness, amplitude, and width conditions. Since there are many spurious peaks in the spectrum generated with $P = 80$ or 100, we add the restriction that harmonic peaks should be aligned consecutively in the lower and middle frequency bands and that the fundamental frequency should be below 250Hz in such spectrum based on the feature of speech signals. Also, we apply a confidence level to the detection result when $P = 40$, which can be either 0 or 1. If the confidence level is 1, we proceed to the next signal frame. Otherwise, we will attempt to detect harmonic peaks in the spectrum with a higher resolution (i.e. $P = 80$ and 100). If no harmonic peaks are detected in these spectra, we will come back to take the result of $P = 40$. Otherwise, we adopt harmonic peaks detected in a spectrum with a higher order. Harmonic peaks detected through the above procedure for some frames of song and speech signals are shown in Figure 3.7, where each detected peak is marked with a vertical line, and P is order of the AR model.

Harmonic peaks are detected once every 100 input samples, and each signal frame contains 512 samples. Therefore, there are overlaps on neighboring frames. The locations of detected peaks are aligned in the temporal order to form the spectral peak tracks. There are also two steps of post-processing applied to the obtained tracks in order to correct misdetections. The first step is called "linking", in which some missing points in the tracks are added according to contextual relations to make these tracks complete. These missing points may result from weak or overlapped harmonic peaks which are difficult to detect. The second step is called "cleaning", which is to remove isolated points in the tracks for the ease of further processing.

Spectrograms and spectral peak tracks estimated with our method for four segments of song and speech signals are illustrated in Figures 3.8 - 3.11. The first segment is female vocal solo without musical instrument accompaniment. There are seven notes sung in the segment as "5-1-6-4-3-1-2". We can see that the pitch and the duration of each note are clearly reflected in the detected peak tracks. Each note lasts for

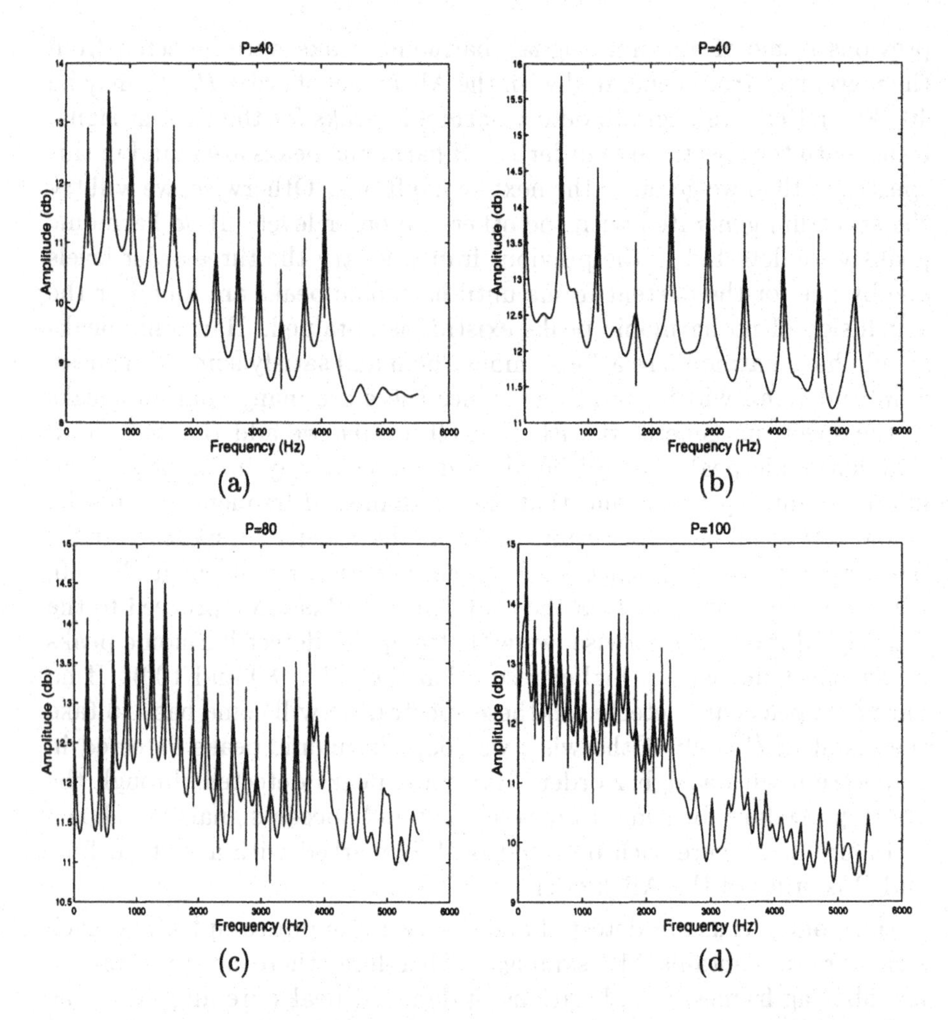

Figure 3.7. Detecting harmonic peaks from power spectrum generated with the AR model parameters for song and speech segments: (a) female song with P=40, (b) female song with P=40, (c) female speech with P=80 and (d) male speech with P=100.

about 0.7-0.8 second. The harmonic tracks range from the fundamental frequency at about 225-400Hz up to 5000Hz, and are in a ripple-like shape. The amplitudes of ripples increase from lower frequency bands to higher frequency bands. The second segment is also female vocal solo but with instrument accompaniment. Some tracks are not so tidy as those in the first segment due to the background noise and some shorter notes. However, we still can find many groups of ripple-shaped tracks, especially that tracks of longer notes are pretty clean at the lower to middle frequency bands. The fundamental frequency is from 550 to

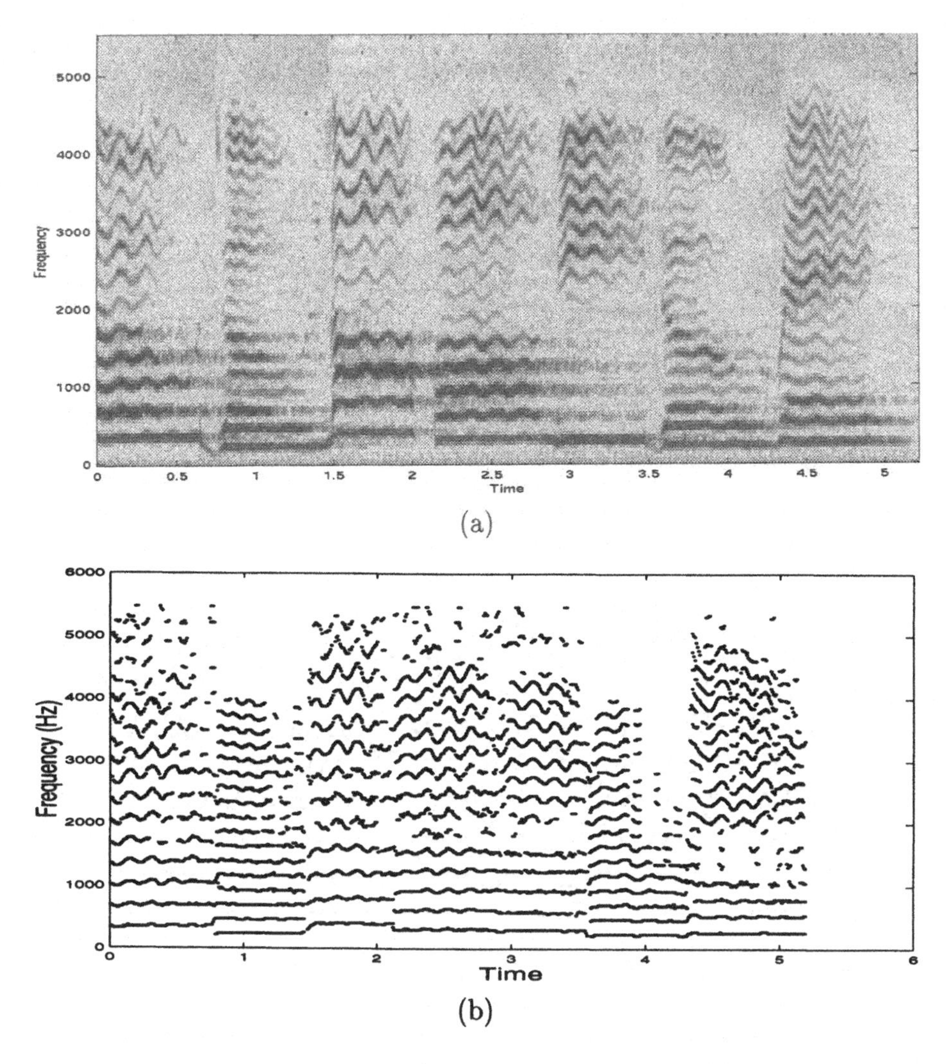

Figure 3.8. The spectrogram and spectral peak tracks of female vocal solo (with no accompanying instrumental sound).

650Hz in this segment. The third segment is female speech having music and other noise in the background. However, the speech signal seems to be dominant in the spectrogram, and the spectral peak tracks are nicely detected despite the interference. The tracks are shorter than those in the song segments with a pitch level of 150-250Hz. The forth segment is a short piece of male speech with noisy background. Again, in spite of the low pitch level (at about 100-130Hz) and background disturbance, most tracks are clearly shown in the picture. The tracks concentrate at the lower to middle parts of the frequency axis.

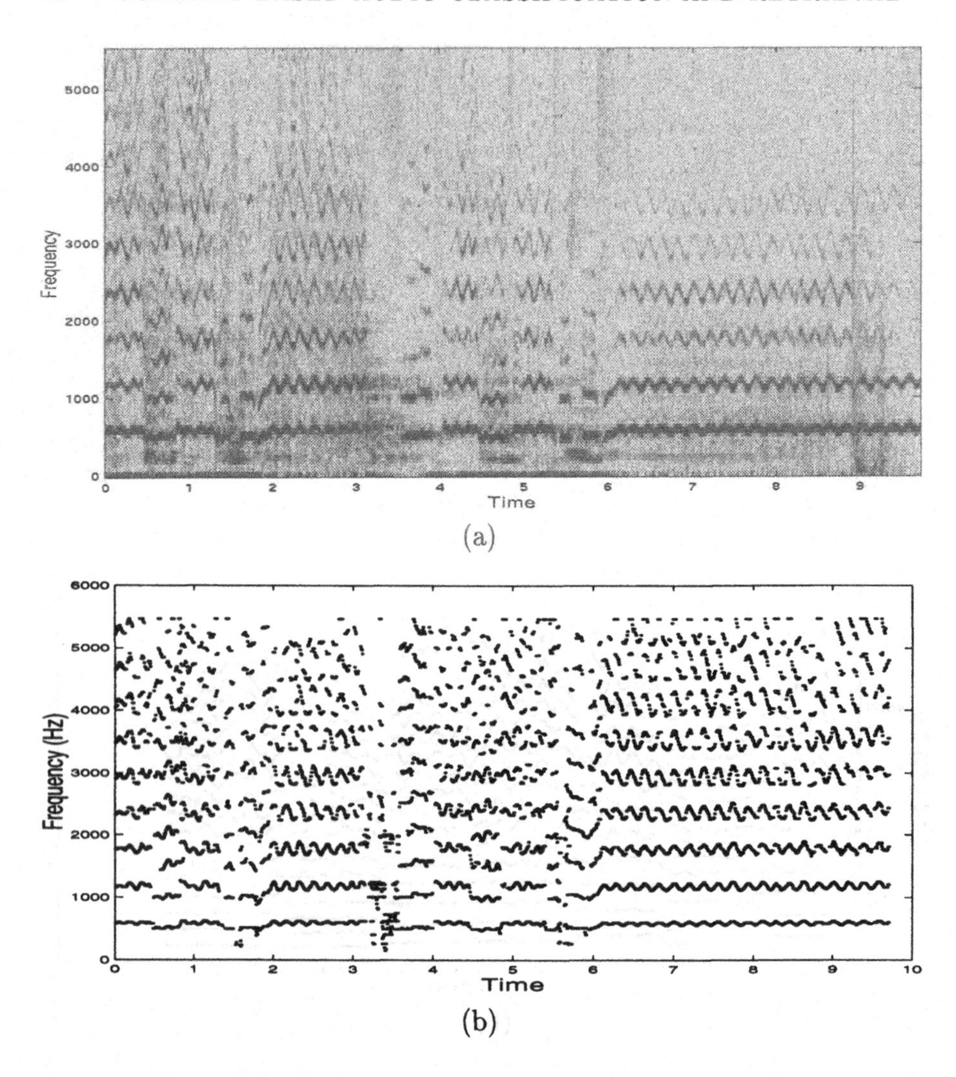

Figure 3.9. The spectrogram and spectral peak tracks of female vocal solo (with accompanying instrumental sounds).

2. AUDIO FEATURES FOR FINE-LEVEL CLASSIFICATION AND RETRIEVAL OF SOUND EFFECTS

2.1 TIMBRE FEATURES

Timbre is generally defined as "the quality which allows one to tell the difference between sounds of the same level and loudness when made by different musical instruments or voices". From the physical point of view, timbre depends primarily upon the spectrum of the stimulus.

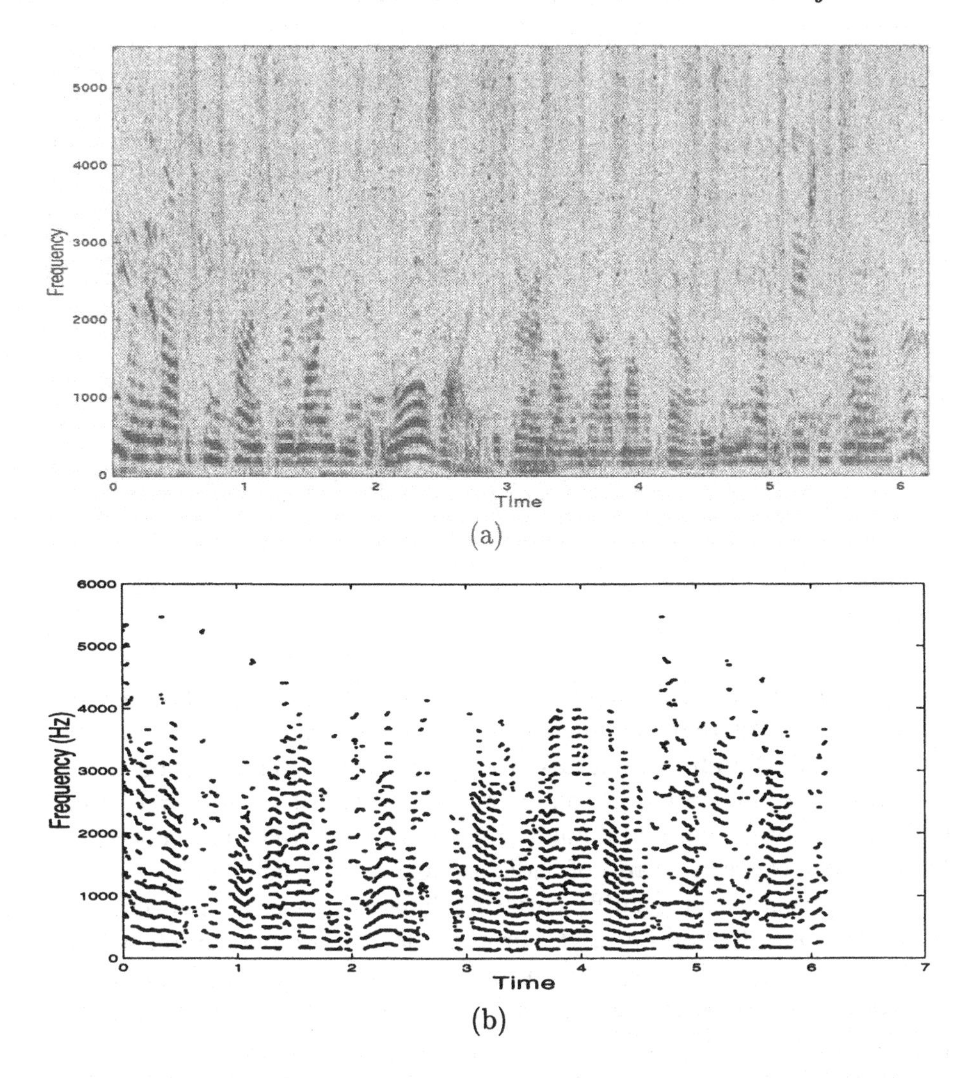

Figure 3.10. The spectrogram and spectral peak tracks of female speech with music and noise in the background.

It also depends upon the waveform, the sound pressure, the frequency location of the spectrum and the temporal characteristics of the stimulus [44]. In music, it is normally believed that timbre is determined by the number and relative strengths of the instrument's partials. However, this is only close to be true [42]. The problem of building physical models for timbre perception has been investigated for a long time in psychology and music analysis without definite answers [45] - [48]. Nevertheless, we may get the conclusion from existing results that the temporal evolution of spectrum of audio signals accounts largely for timbre perception.

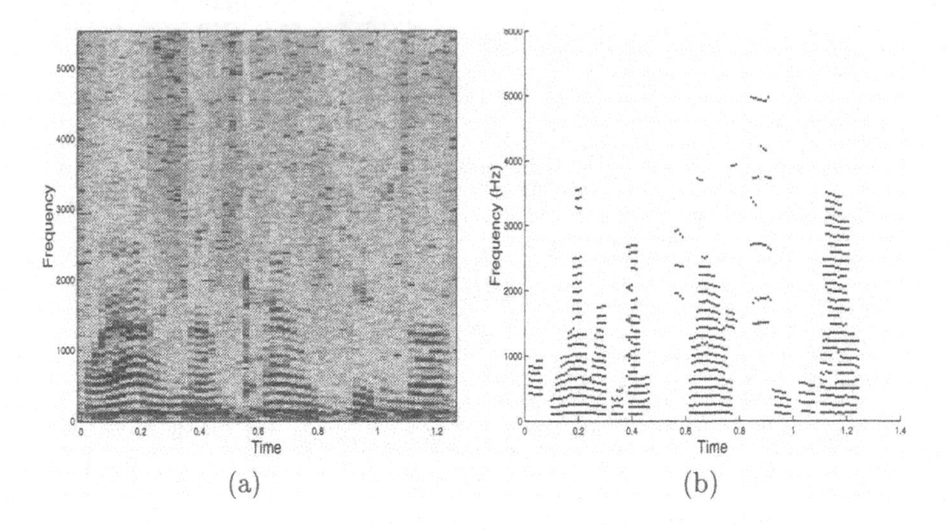

Figure 3.11. The spectrogram and spectral peak tracks of male speech with noisy background.

We observed a large amount of environmental sounds of different kinds, and found that the timbre patterns were well reflected in the spectrograms of audio waveforms. Here, we extend timbre from a term originally used for harmonic sound (music and voice) to the perception of environmental sound, and analyze it on the time-frequency representation (such as spectrogram) of audio signals. We consider timbre as the most important feature in differentiating different classes of environmental sounds, and to build a model properly for timbre perception based on the spectrogram is one major problem in this research. Figure 3.12 illustrates the spectrograms of several environmental sounds. For example, the sound shown in Figure 3.12(b) includes two sorts of timbres: the bird's cry (of higher frequency) and the river flow sound in the background (in lower frequency bands), which can be clearly observed from the spectrogram. The sound of thunder contains intense components of lower frequencies. The sounds of footstep and applause cover a broader range of frequency bands, but they also have very different change patterns of the frequency distribution over the time.

In this work, we use the hidden Markov model (HMM) to describe the temporal evolution of the spectral envelope of environmental sounds for the purpose of modeling timbre perception. The temporal evolution pattern is contained in the transition and duration parameters of HMM, and the observation vector in HMM at each instant is the spectral envelope of the audio signal frame around that time point. A key problem here is how to extract the feature vector to represent the spectral en-

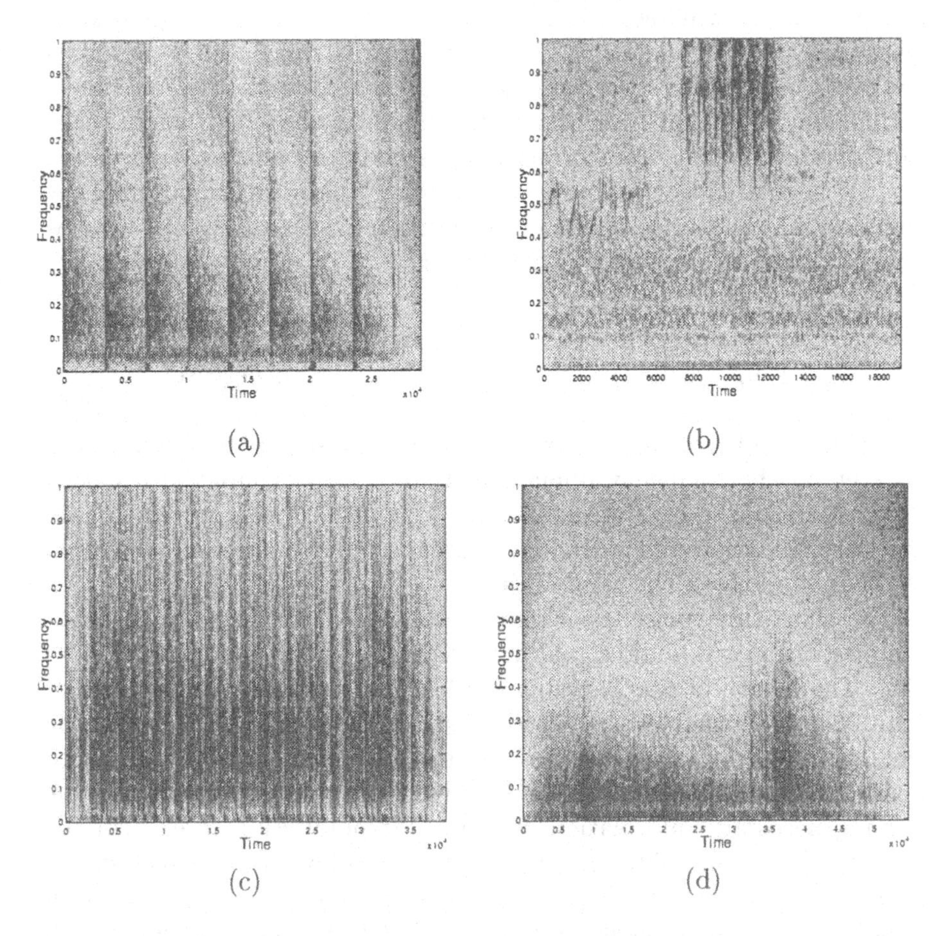

Figure 3.12. The spectrograms of environmental sounds: (a) footstep, (b) bird-river, (c) applause and (d) thunder.

velope. We choose to use a smoothed version of the short-time power spectrum of audio signal.

We compared three methods for estimating the short-time power spectrum of audio signals. The first one is the "direct method", in which the spectrum is computed by applying FFT directly on the sound waveform as below:

$$\hat{S}_{direct} = \log |FFT(x(m)w(n-m))|, \tag{3.5}$$

where

$$w(n) = \begin{cases} 0.5(1-\cos(2\pi \frac{n}{N-1})), & 0 \leq n \leq N-1, \\ 0, & \text{otherwise,} \end{cases}$$

is the Hanning window, which is chosen for its relatively small sidelobes and fast attenuation. The second one is called the "AR model method", where the spectrum $\hat{S}_{AR}$ is generated with the AR model parameters estimated from the audio signal as described before. The third method is the "autocorrelation method", in which the spectrum $\hat{S}_{auto}$ is calculated as the Fourier transform of the autocorrelation of the audio signal. The autocorrelation is multiplied with a Papoulis window before the transform, which is defined as:

$$w(n) = \begin{cases} \frac{1}{\pi}|\sin(\frac{2\pi n}{N})| - (1 - \frac{2|n-\frac{N}{2}|}{N})\cos(\frac{2\pi n}{N}), & 0 \leq n \leq M-1, \\ 0, & \text{otherwise,} \end{cases} \tag{3.6}$$

where M is width of the window. The Papoulis window function is non-negative in the frequency domain so that it may ensure the resulting power spectrum to be always positive. Also, it has been proved to have the least deviation and variance in the spectrum estimation compared to other windows of the same width.

The short-time power spectrum is computed once every 100 input samples, with each audio signal frame containing 150 samples, i.e. $N = 150$. The length of the Papoulis window M is normally much smaller than N and determines the degree of smoothness of $\hat{S}_{auto}$. That is, the smaller M is, the smoother the obtained spectrum will be. Trying to maintain a low dimension of the feature vector while at the same time keeping necessary information, we take 128-point FFT in each of the three methods, thus obtaining a feature vector of 65 dimensions (i.e. the logarithm of the amplitude spectrum at each frequency sample between 0 and π). For audio signals sampled at 11025Hz, there are about 110 feature vectors obtained per second for each sound.

Shown in Figure 3.13 are examples of the estimated power spectrum for a certain frame of environmental sound. In both pictures, plotted with dashdot lines are $\hat{S}_{direct}$. The curve in the dashed line in (a) is $\hat{S}_{AR}$ (where the order of AR model is 10), and the one in the dashed line in (b) is $\hat{S}_{auto}$ (where $M = 63$). Theoretically, $\hat{S}_{direct}$ has the least deviation and the best frequency resolution. The "direct method" also has the least computational cost. However, $\hat{S}_{direct}$ as well has the largest variance, as seen from the dramatic changes in the curve. The AR model generated spectrum is very smooth. Also, the "AR model method" has the least number of parameters. That is, even though the dimension of feature vector is 65, we only need to save the AR model parameters which are of a much smaller dimension. But, as an all-pole model, it can only follow peaks in the spectrum while troughs are not properly represented. As shown in Figure 3.13(a), the amplitude level of $\hat{S}_{AR}$ is generally higher

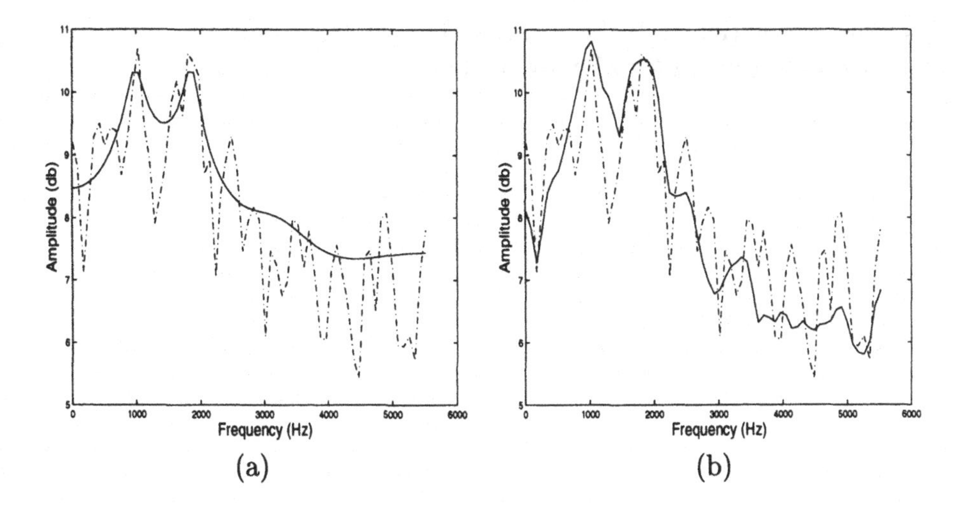

Figure 3.13. Short-time power spectrum of environmental sound estimated with different methods: (a) "AR model method" vs. "direct method" and (b) "autocorrelation method" vs. "direct method".

than that of $\hat{S}_{direct}$. When $M = N$, $\hat{S}_{auto}$ is the same as $\hat{S}_{direct}$. When $M < N$, it is a smoothed version of $\hat{S}_{direct}$. The smoothness is obtained at the cost of loosing the frequency resolution. Nevertheless, $\hat{S}_{auto}$ keeps the general trend of $\hat{S}_{direct}$. On the other hand, the complexity of the "autocorrelation method" is higher due to the computation of the short-time autocorrelation. When used in the classification and retrieval of sound effects with HMM, it is found that $\hat{S}_{auto}$ has the best results. If the computational cost is critical, $\hat{S}_{direct}$ may be used. The corresponding results are fine because the high variances in $\hat{S}_{direct}$ are cancelled out in the averaging process of building model parameters. $\hat{S}_{AR}$ normally does not have a satisfactory performance in these applications.

2.2 RHYTHM FEATURES

Rhythm is a term originally defined for speech and music. It is the quality of happening at regular periods of time. Here, we extend it to environmental sounds to represent the change pattern of timbre and energy in a sound clip. One example is shown in Figure 3.12(a), where the rhythm of footstep is a significant feature of the sound. Other sounds in which rhythm plays an important role in the perception include clock tick, telegraph machine, pager, door knock, etc. For instance, all sounds of knocking have the quasi-periodic change of energy level, even though the timbre feature may be different from one sound clip to another (depending on how you knock and on what you knock). In this work, the rhythm information is to be reflected by the transition and duration parameters in the hidden Markov model.

Chapter 4

GENERIC AUDIO DATA SEGMENTATION AND INDEXING

In the coarse-level segmentation and indexing stage, audio data are segmented and classified into basic audio types, based on morphological and statistical analysis of the temporal curves of the short-time energy function, the short-time average zero-crossing rate, and the short-time fundamental frequency, as well as the spectral peak tracks of audio signals. Threshold-based heuristical rules are derived empirically to guide the classification procedures. Therefore, the approach is completely generic and model-free, which can be applied under any circumstances. An illustration of the scheme is shown in Figure 4.1.

1. DETECTION OF SEGMENT BOUNDARIES

For on-line segmentation of audiovisual data, the short-time energy function, short-time average zero-crossing rate, and short-time fundamental frequency are computed on the fly with incoming audio data. Whenever there is an abrupt change detected in any of these three features, a segment boundary is set. In the temporal curve of each feature, there are two adjoining sliding windows installed with the average amplitude computed within each window (as illustrated in Figure 4.2). The sliding windows proceed together with newly computed feature values, and the average amplitudes $Ave(w1)$ and $Ave(w2)$ are updated. We compare these two values, and whenever there is a big difference between them, an abrupt change is claimed to be detected at the common edge of the two windows (i.e. the point E). We choose the length of each window to be 100 feature samples, which equals to about one second in time.

Examples of boundary detection from temporal curves of the short-time energy function and the short-time fundamental frequency are shown

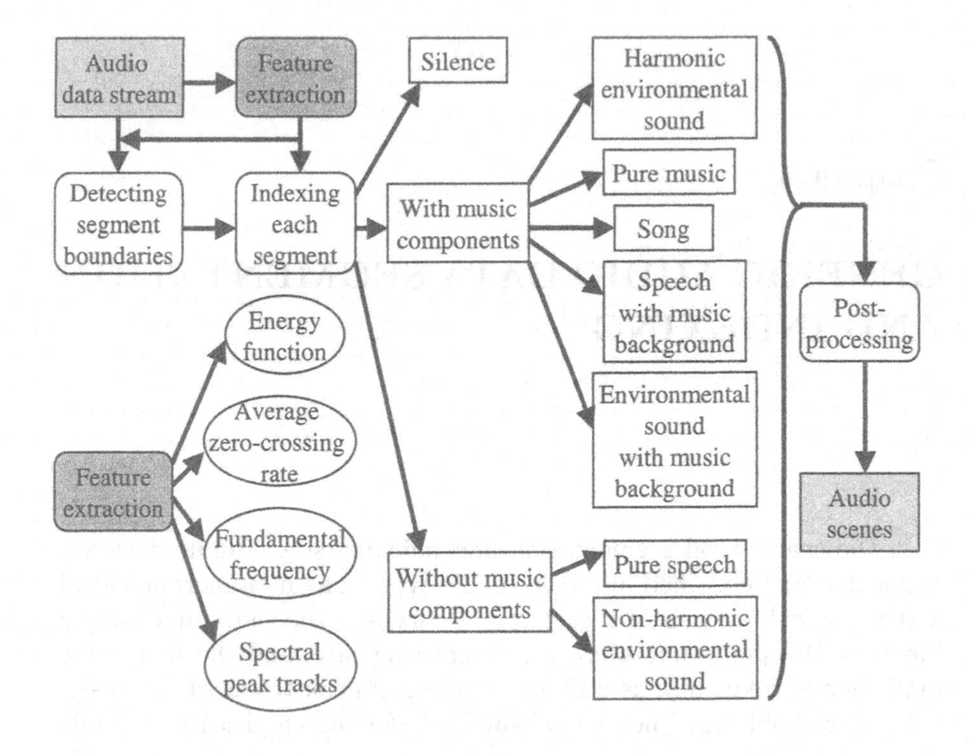

Figure 4.1. Automatic segmentation and indexing of the audio data stream into semantic audio scenes.

in Figure 4.3. We can see that because the temporal evolution pattern and range of amplitudes of the short-time features are different for speech, music, environmental sound, etc. dramatic changes can be detected from these features at the boundaries of different audio types.

2. CLASSIFICATION OF EACH SEGMENT

After segment boundaries are detected, each segment is classified into one of the basic audio types through the following steps.

2.1 DETECTING SILENCE

The first step is to check whether the audio segment is silence or not. We define "silence" to be a segment of imperceptible audio, including unnoticeable noise and very short clicks. The normal way to detect silence is by energy thresholding. However, we have found that the energy level of some noise pieces is not lower than that of some music pieces. The reason that we can hear the music while may not notice the noise is that the frequency-level of noise is much lower. Thus, we use both

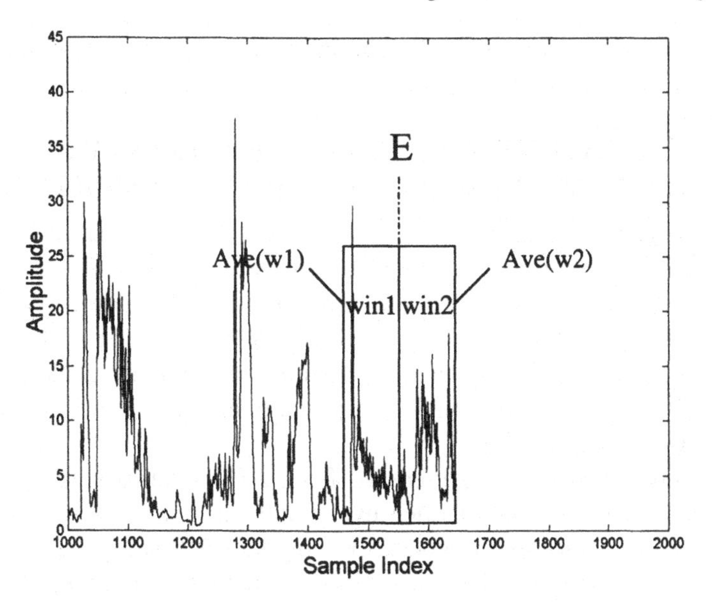

Figure 4.2. Setting sliding windows in the temporal curve of an audio feature for boundary detection.

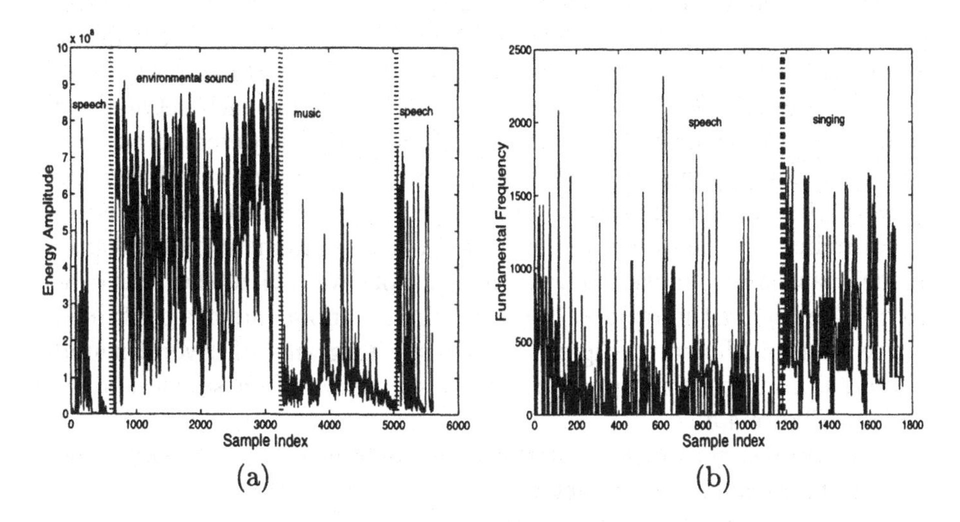

Figure 4.3. Boundary detection in the temporal curves of (a) short-time energy function and (b) short-time fundamental frequency.

energy and ZCR measures to detect silence. If the short-time energy function is continuously lower than a certain set of thresholds (there may be durations in which the energy is higher than the threshold, but the durations should be short enough and far apart from each other), or if most values of the short-time average zero-crossing rate in the segment are lower than certain set of thresholds, then the segment is indexed as "silence".

2.2 SEPARATING SOUNDS INTO WITH AND WITHOUT MUSIC COMPONENTS

As observed from movies and video programs, music is an important type of audio component frequently appearing, either alone or as the background of speech or environmental sounds. Therefore, we first separate the audio segments into two categories, i.e. with or without music components, mainly by detecting continuous frequency peaks from the power spectrum. That is, music components are characterized by stable power spectrum peaks [49].

The power spectrum is generated from AR model parameters of the order 40, and is calculated once every 400 input samples. The signal frame used to compute the spectrum contains 512 samples, and a 512-point FFT is used. If there are peaks detected in consecutive power spectra which stay at about the same frequency level for a certain period of time, this period of time is indexed as having music components. In order to avoid the influence from speech harmonic peaks or low frequency noise, we only consider peaks above 500Hz, because most music components are in this range. Signal frames below a certain energy level are also ignored. An index sequence is generated for each segment of sound, i.e. the index value is set to 1 if the sound is detected as having music components at that instant and to 0, otherwise. The ratio between the number of zeros in the index sequence and the total number of indices in the sequence can thus be a measurement of the sound segment as having music components or not (we call it the "zero ratio"). The higher the ratio is, the less music components are contained in the sound. Shown in Figure 4.4 are spectrograms of several sound segments, and the corresponding index sequences of these sounds are given in Figure 4.5.

We examined zero ratios of indices for different types of sounds, and the results are summarized below.

1. Speech:
   Although the speech signal contains many harmonic components, the frequency peaks change faster and last for a shorter time than those of music. Zero ratios for speech segments are normally above 0.95. Misdetections are mainly due to the lengthened pronunciations or spurious frequency peaks.

2. Environmental Sound:
   Harmonic and stable environmental sounds are all indexed as hav-

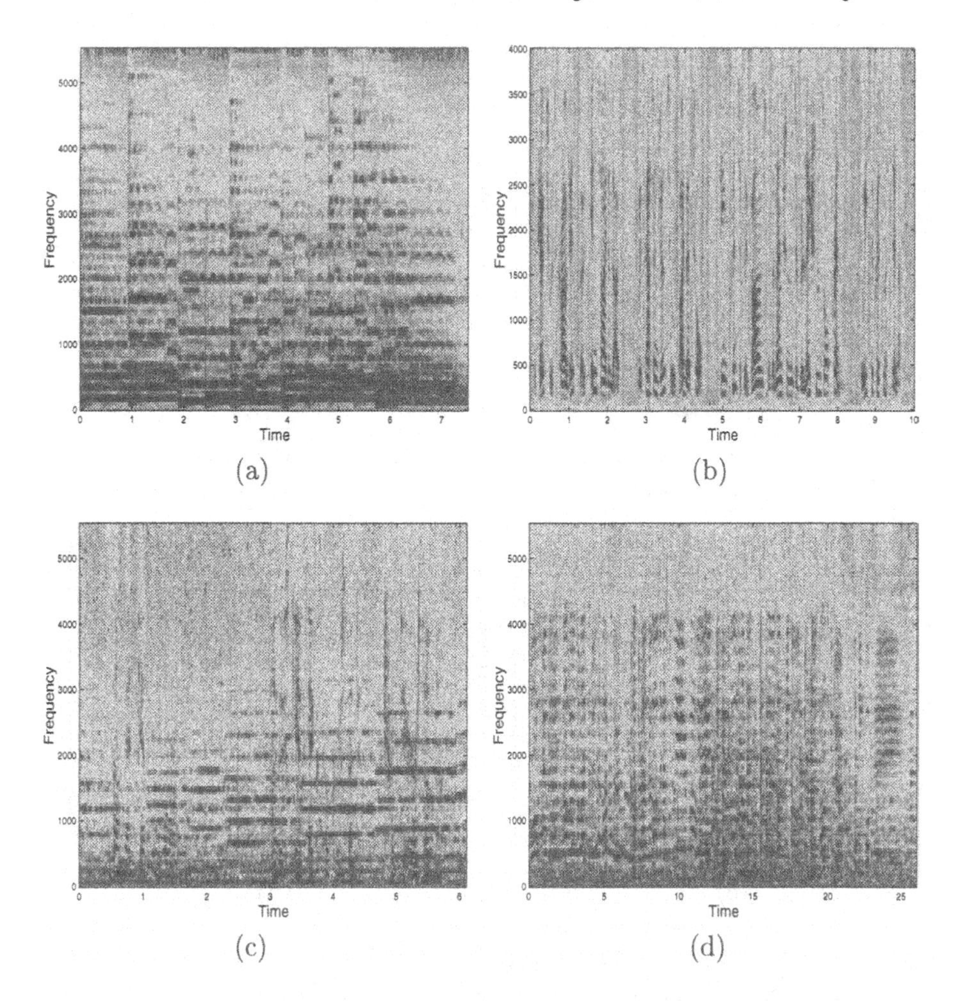

Figure 4.4. The spectrograms of sound segments: (a) pure music, (b) pure speech, (c) speech with music background and (d) song.

ing music components, while non-harmonic sounds are all indexed as not having music components. However, there are some exceptional cases in between, e.g. some harmonic and non-harmonic mixed environmental sounds, for which we have to add rules in the program to properly place them.

3. Pure Music:
   Zero ratios for all pure music segments are below 0.3. Indexing errors normally come from short notes, low volume or low frequency parts, non-harmonic components, and the intermissions between two notes.

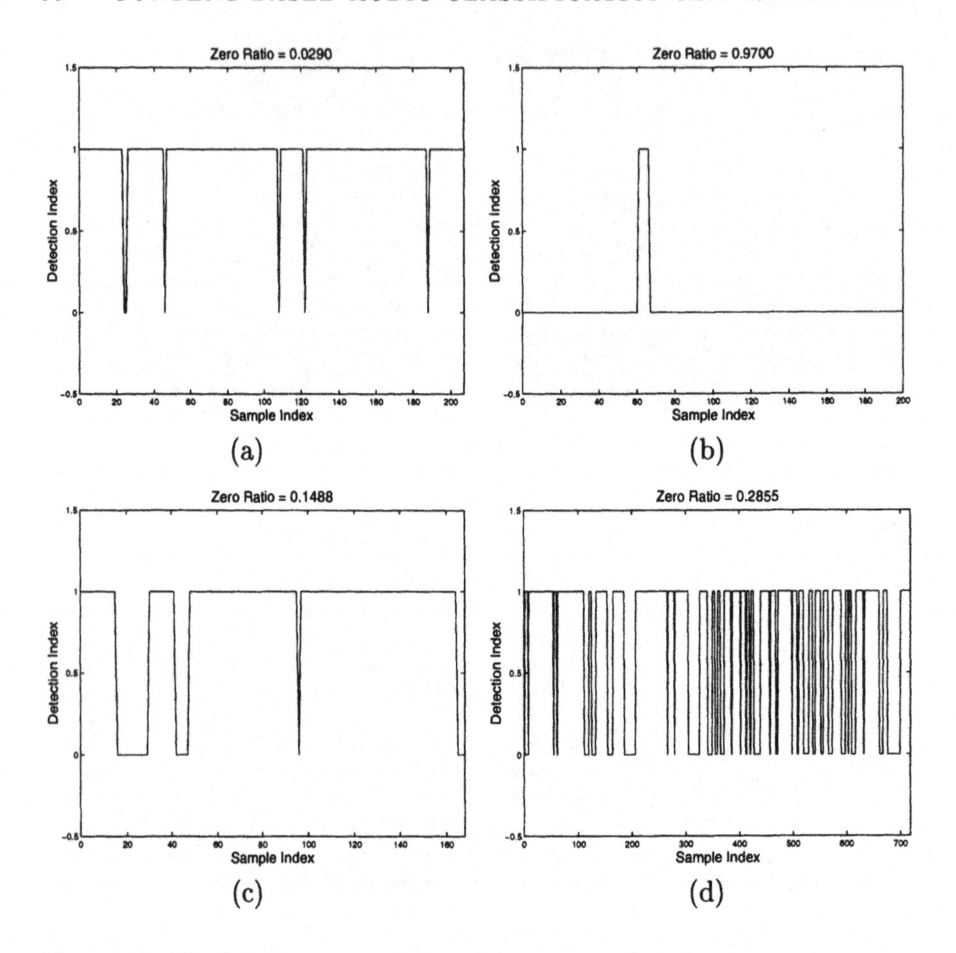

Figure 4.5. The index sequences for music component detection in sound segments of: (a) pure music, (b) pure speech, (c) speech with music background and (4) song (with zero ratio value above each picture).

4. Song:
   Most song segments have zero ratios below 0.5. Those parts not detected as having music components result from: those peak tracks that shaped like ripples instead of lines when the note is long, the intermissions between notes (due to breaths of the singer), low volume and/or low frequency sounds. When the ripple-shaped peak tracks are detected and indexed as music components, zero ratios for songs are significantly reduced.

5. Speech with Music Background:
   When the speech is strong, the background music is normally hidden and can not be detected. However, music components can be detected

in the intermission periods in speech or when music becomes stronger. We make the distinction of the following two cases. For the first case, when music is stronger or there are many intermissions in speech so that music is a prominent part of the sound, the zero ratios are below 0.6. For the second case, when music is weak while speech is strong and continuous, speech is the major component and music may be ignored. Zero ratios are higher than 0.8 in such a case.

Therefore, based on a threshold for the zero ratio at about 0.7 together with some other rules, audio segments can be separated into two categories as desired. The first category contains harmonic and stable environmental sound, pure music, song, speech with the music background, and environmental sound with the music background. For the second category, there are pure speech and non-harmonic environmental sounds. Further classification will be done within each category.

2.3 DETECTING HARMONIC ENVIRONMENTAL SOUNDS

The next step is to separate out environmental sounds which are harmonic and stable. The temporal curve of the short-time fundamental frequency is checked. If most parts of the curve are harmonic, and the fundamental frequency is fixed at one particular value, then the segment is indexed as "harmonic and unchanged". A typical example of this type is the sound of touch-tone. If the fundamental frequency of a sound clip changes over time but only with several values, it is indexed as "harmonic and stable". Examples of this type include the sounds of the doorbell and the pager. This classification step is performed here as a screening process for harmonic environmental sounds, so that they will not confuse the differentiation of music. It is also the basis for further fine-level classification of harmonic environmental audio.

2.4 DISTINGUISHING PURE MUSIC

Pure music is distinguished based on the average zero-crossing rate and the fundamental frequency properties. Four aspects are checked. They are the degree of being harmonic, the degree of the fundamental frequency's concentration on certain values during a period of time, the variance of the average zero-crossing rates, and the range of amplitudes of the average zero-crossing rates. For each aspect, there is one empirical threshold set and a decision value defined. If the threshold is satisfied, the decision value is set to 1; otherwise, it is set to a fraction between 0 and 1 according to the distance to the threshold. The four decision values are averaged with predetermined weights to derive a total probability of the audio segment's being pure music. For a segment to be indexed as "pure music", this probability should be above a certain threshold and at least three of the decision values should be above 0.5.

2.5 DISTINGUISHING SONGS

Up to this step, what are left in the first category include the sound segments of song, speech with the music background and environmental sound with the music background. We extract the spectral peak tracks for these segments, and differentiate the three audio types based on the analysis of these tracks.

Songs may be characterized by one of the three features: ripple-shaped harmonic peak tracks (due to the vibration of vocal chords), tracks which are of a longer duration compared to those in speech, and tracks which have a fundamental frequency higher than 300Hz. These features form the basis for distinguishing song segments. The groups of tracks are checked whether any of these three features is matched. The segment will be indexed as "song" if either the sum of durations in which the harmonic peak tracks satisfy one of the features gets to a certain amount, or its comparison to the total length of the segment reaches a certain ratio. The ripple-shaped tracks are detected by taking the first-order difference of the track and checking the pattern of the resulted sequence. It is worthwhile to point out that some musical instruments such as the violin may also generate ripple-shaped peak tracks, which are normally at higher frequency bands and are of lower amplitudes.

2.6 SEPARATING SPEECH WITH MUSIC BACKGROUND AND ENVIRONMENTAL SOUND WITH MUSIC BACKGROUND

According to [29], "when sounds with peaked spectra are mixed, energy from one or other source generally dominates each channel". Therefore, even though there is music in the background, as long as the speech is strong (i.e. the pronunciations are clear and loud enough for human perception), the harmonic peak tracks of the speech signal can be detected in spite of the existence of music components. We check the groups of tracks to see whether they concentrate in the lower to middle frequency bands (with the fundamental frequency between 100 to 300 Hz) and have lengths within a certain range. If there are durations in which the spectral peak tracks satisfy these criteria, the segment is indexed as "speech with music background".

Then, what left in the first category is the segment which has music components but does not meet the criteria for any of the above audio types. It will be indexed as "environmental sound with music background". Examples for sound segments of these two types are shown in Figure 4.6, where the music background as well as the differences between speech and environmental sound (applause and laugh in this example) in the spectral domain are obvious.

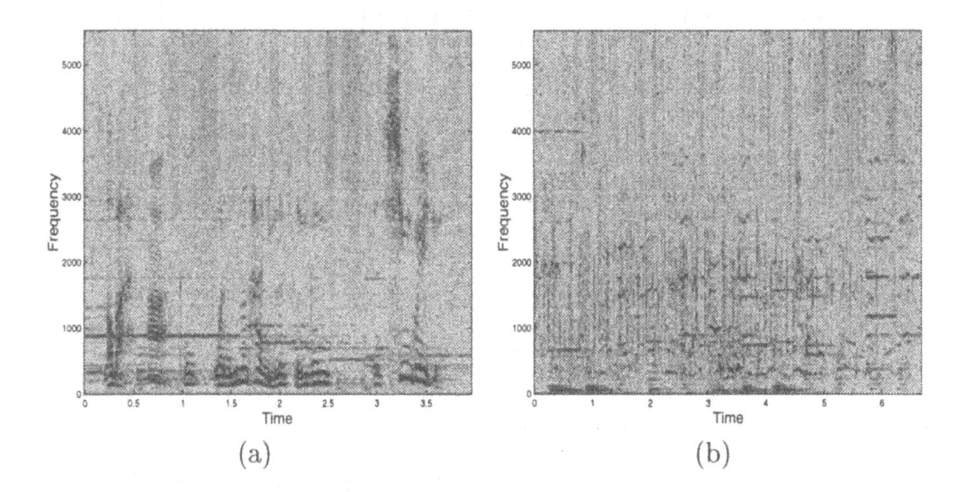

Figure 4.6. The spectrograms of (a) speech with music background segment and (b) environmental sound (applause and laugh) with music background segment.

2.7 DISTINGUISHING PURE SPEECH

When distinguishing pure speech, five aspects of conditions are checked. The first aspect is the relation between the temporal curves of the average zero-crossing rate and the energy function. In speech segments, the ZCR curve has peaks for unvoiced components and troughs for voiced components, while the energy curve has peaks for voiced components and troughs for unvoiced components. Thus, there is a compensative relation between them. One example is shown in Figure 4.7. We clip both ZCR and energy curves at one third of the maximum amplitude and remove the lower parts, so that only peaks of the two curves will remain. Then, the inner product of the two residual curves is calculated. This product is normally near to zero for speech segments because the peaks appear at different parts in the two curves, while it is much larger for other types of audio.

The second aspect is the shape of ZCR curve. For speech, the ZCR curve has a stable and low baseline with peaks above it. We define the baseline to be the linking line of lowest points of troughs in the ZCR curve. The mean and variance of the baseline are calculated. The

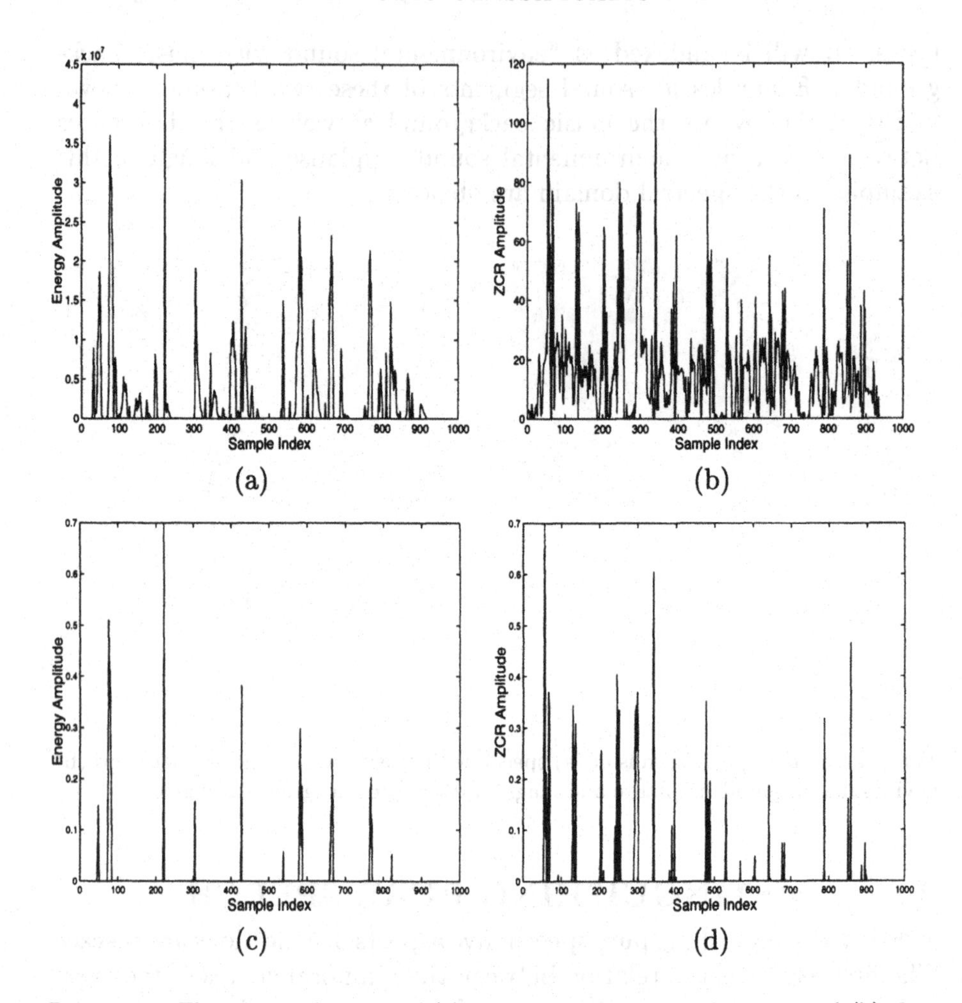

Figure 4.7. The temporal curves of (a) the short-time energy function and (b) the short-time average zero-crossing rate of a speech segment. (c) and (d) are the residual curves of (a) and (b), respectively, after clipping and normalization.

parameters (amplitude, width, and sharpness) and the frequency of appearance of the peaks are also considered. The third and fourth aspects are the variance and the range of amplitudes of the ZCR curve, respectively. Contrary to music segments where the variance and the range of amplitudes are normally lower than certain thresholds, a typical speech segment has a variance and a range of amplitudes that are higher than certain thresholds.

The fifth aspect is about the property of the short-time fundamental frequency. As voiced components are harmonic and unvoiced components are non-harmonic, speech has a percentage of harmony within a certain range. There is also a relation between the fundamental frequency curve and the energy curve. That is, the harmonic parts in the SFuF curve correspond to peaks in the energy curve while the zero parts in the SFuF curve correspond to troughs in the energy curve. A decision value, which is a fraction between 0 and 1, is defined for each of the five aspects. The weighted average of these decision values represent the possibility of the segment's being speech. When the possibility is above a certain threshold and at least three of the decision values are above 0.5, the segment is indexed as "pure speech".

2.8 CLASSIFYING NON-HARMONIC ENVIRONMENTAL SOUNDS

The last step is to classify what is left in the second category into one type of non-harmonic environmental sounds as the following. *(1)* If either the energy function curve or the average zero-crossing rate curve has peaks which have approximately equal intervals between neighboring peaks, the segment is indexed as *"periodic or quasi-periodic"*. Examples for this type include sounds of the clock tick (as shown in Figure 4.8(a)) and the regular footstep. This is a beginning of rhythm analysis. More complicated rhythm analysis will be done in the fine-level classification of environmental sound. *(2)* If the percentage of harmonic parts in the fundamental frequency curve is within a certain range (lower than the threshold for music, but higher than the threshold for non-harmonic sound), the segment is indexed as *"harmonic and non-harmonic mixed"*. For example, the sound of train horn, which is harmonic, appears with a non-harmonic background. Also, the sound of cough consists of both harmonic and non-harmonic components. The spectrograms of these sounds are shown in Figure 4.9. *(3)* If the frequency centroid (denoted by the average zero-crossing rate value) is within a relatively small range compared to the absolute range of the frequency distribution, the segment is indexed as *"non-harmonic and stable"*. One example is the sound of birds' cry, which is non-harmonic but its ZCR curve is concentrated within the range of 80-120 (as illustrated in Figure 4.8(b)). *(4)* Finally, if the segment does not satisfy any of the above conditions, it is indexed as *"non-harmonic and irregular"*. Many environmental sounds belong to this type, such as the sounds of thunder, earthquake and fire.

3. POST-PROCESSING

The post-processing step is to reduce possible segmentation and classification errors. We have adjusted the segmentation algorithm to be sensitive enough to detect all abrupt changes. Thus, it is possible that

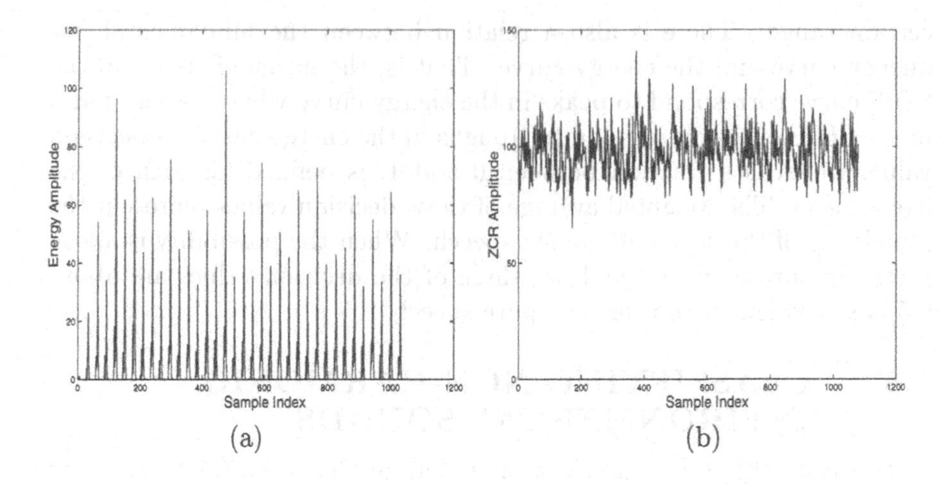

Figure 4.8. **The temporal curves of: (a) short-time energy function of clock tick and (b) short-time average zero-crossing rate of birds' sound.**

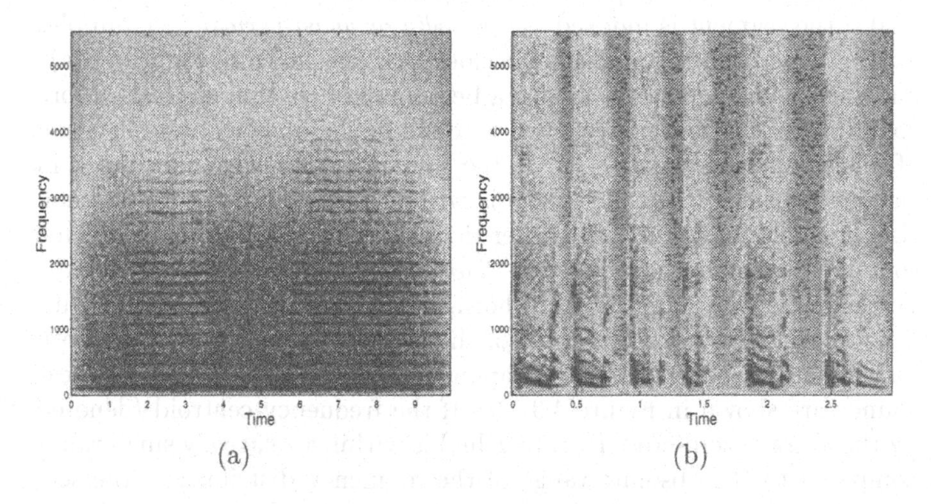

Figure 4.9. **The spectrograms of the sounds of (a) trainhorn and (b) cough.**

one continuous scene is broken into several segments. For example, one music piece may be broken into several segments due to abrupt changes in the energy curve, and some small segments may even be misclassified as "harmonic and stable environmental sound" because of the unchanged tune in the segment. Through post-processing, these segments are to be combined to other segments and are reindexed based on their contextual relations.

Here are some examples of heuristic rules used in the post-processing step: if a "silence" segment is shorter than two second and the two segments prior and next to it have the same index, the three segments are merged into one and indexed as the same as the first segment; if a "harmonic and fixed" or "harmonic and stable" environmental sound segment is shorter than five second and is next to a "pure music" or "song" segment, it is merged into that segment; if a "harmonic and non-harmonic mixed environmental sound" is shorter than two second and is between two segments of "speech with music background" or "environmental sound with music background", the three segments are merged into one and indexed according to the first segment.

Chapter 5

SOUND EFFECTS CLASSIFICATION AND RETRIEVAL

The core of the fine-level classification and retrieval of environmental sound (including sound effects) is to build the hidden Markov model (HMM) for each class or clip of sound(s). Currently, two types of information are contained in the model, i.e. timbre and rhythm. Each kind of timbre is modeled as one state of the HMM, and represented with the Gaussian mixture density. The rhythm information is included in the transition parameters of HMM. For sound effects classification, HMM serves as the classifier; while for sound effects retrieval, it is the similarity measure.

There are in general two reasons that we choose the hidden Markov model as the major tool in this part of the work. First, the transition and duration parameters in HMM may properly reveal the evolution of features over time, which is very important in modeling audio perception. For example, with HMM, it is convenient to differentiate sounds with stable spectrum, repetitive spectrum, or randomly changed spectrum. Second, there are many kinds of variations of the HMM as well as experiences of using them which are developed in the Speech Recognition researches. This makes HMM a mature technique to be applied in this research.

1. HIDDEN MARKOV MODEL AND GAUSSIAN MIXTURE MODEL

The hidden Markov model (HMM) and the Gaussian mixture model (GMM) are powerful statistical tools widely used in pattern recognition. Especially, the HMM has been developed extensively in Speech Recognition systems over the last three decades [50] - [52], while the GMM has been used successfully for Speaker Identification in recent years [53], [54]. In this work, they are used to characterize the timbres and their change pattern(s) in one sound clip or a class of sounds. GMM can be viewed as one component of HMM under certain circumstances.

1.1 THE GAUSSIAN MIXTURE MODEL

A Gaussian mixture density is a weighted sum of M component densities, as given by the following [53]:

$$p(\vec{x}|\lambda) = \sum_{i=1}^{M} p_i b_i(\vec{x}), \tag{5.1}$$

where $\vec{x}$ is a D-dimensional random vector, $b_i(\vec{x})$, $i = 1, \ldots, M$, are the component densities, and p_i, $i = 1, \ldots, M$, are the mixture weights. Each component density is a D-variate Gaussian function of the form

$$b_i(\vec{x}) = \frac{1}{(2\pi)^{D/2}|\Sigma_i|^{1/2}} \exp\{-\frac{1}{2}(\vec{x} - \vec{\mu}_i)' \Sigma_i^{-1} (\vec{x} - \vec{\mu}_i)\}, \tag{5.2}$$

with mean vector $\vec{\mu}_i$ and covariance matrix Σ_i. The mixture weights have to satisfy the constraint $\sum_{i=1}^{M} p_i = 1$.

The complete Gaussian mixture density is parameterized by the mean vectors, the covariance matrices and the mixture weights from all component densities. These parameters are collectively represented by

$$\lambda = \{p_i, \vec{\mu}_i, \Sigma_i\}, \quad i = 1, \ldots, M. \tag{5.3}$$

In the training process, the maximum likelihood (ML) estimation is adopted to determine model parameters which maximize the likelihood of GMM given the training data. For a sequence of T training vectors, $X = \{\vec{x}_1, \ldots, \vec{x}_T\}$, the GMM likelihood can be written as

$$p(X|\lambda) = \prod_{t=1}^{T} p(\vec{x}_t|\lambda). \tag{5.4}$$

The ML estimates of the GMM parameters are obtained iteratively by using the expectation-maximization (EM) algorithm. At each iteration, the reestimation formulas are as follow, which guarantee a monotonic increase in the likelihood value.

- Mixture weight update:

$$\overline{p}_i = \frac{1}{T} \sum_{t=1}^{T} p(i|\vec{x}_t, \lambda). \tag{5.5}$$

- Mean vector update:

$$\overline{\vec{\mu_i}} = \frac{\sum_{t=1}^{T} p(i|\vec{x}_t, \lambda)\vec{x}_t}{\sum_{t=1}^{T} p(i|\vec{x}_t, \lambda)}. \tag{5.6}$$

- Covariance matrix update:

$$\overline{\Sigma_i} = \frac{\sum_{t=1}^{T} p(i|\vec{x}_t, \lambda)(\vec{x}_t - \overline{\vec{\mu_i}})(\vec{x}_t - \overline{\vec{\mu_i}})'}{\sum_{t=1}^{T} p(i|\vec{x}_t, \lambda)}. \tag{5.7}$$

The *a posteriori* probability for the ith mixture is given by

$$p(i|\vec{x}_t, \lambda) = \frac{p_i b_i(\vec{x}_t)}{\Sigma_{k=1}^{M} p_k b_k(\vec{x}_t)}. \tag{5.8}$$

1.2 THE HIDDEN MARKOV MODEL

A hidden Markov model for discrete symbol observations is characterized by the following parameters [50]:

1. N, the number of states in the model. We label the individual states as $\{1, 2, \ldots, N\}$, and denote the state at time t as q_t.
2. M, the number of distinct observation symbols in all states, i.e. the discrete alphabet size. We denote the individual symbols as $V = \{\mathbf{v}_1, \mathbf{v}_2, \ldots, \mathbf{v}_M\}$.
3. The state-transition probability distribution $A = \{a_{ij}\}$ where

$$a_{ij} = P[q_{t+1} = j | q_t = i], \;\; 1 \le i, \; j \le N. \tag{5.9}$$

4. The observation symbol probability distribution $B = \{b_j(k)\}$, in which

$$b_j(k) = P[\mathbf{x}_t = \mathbf{v}_k | q_t = j], \;\; 1 \le k \le M, \tag{5.10}$$

   defines the symbol distribution in state j, $j = 1, 2, \ldots, N$.

5. The initial state distribution $\pi = \{\pi_i\}$ in which

$$\pi_i = P[q_1 = i], \;\; 1 \le i \le N. \tag{5.11}$$

Thus, a complete specification of an HMM includes two model parameters, N and M, the observation symbols, and the three sets of probability measures A, B, and π. We use the compact notation

$$\lambda = (A, B, \pi) \tag{5.12}$$

to indicate the complete parameter set of the model. It is used to define a probability measure for the observation sequence $\mathbf{X}$, i.e. $P(\mathbf{X}|\lambda)$, which can be calculated according to a forward procedure as defined below.

Consider the forward variable $\alpha_t(i)$ defined as

$$\alpha_t(i) = P(\mathbf{x}_1\mathbf{x}_2\ldots\mathbf{x}_t, q_t = i|\lambda), \tag{5.13}$$

which is the probability of the partial observation sequence $\mathbf{x}_1, \mathbf{x}_2, \ldots, \mathbf{x}_t$, and state i at time t, given the model λ. We can solve for $\alpha_t(i)$ inductively as follows.

1. Initialization

$$\alpha_1(i) = \pi_i b_i(\mathbf{x}_1), \quad 1 \leq i \leq N. \tag{5.14}$$

2. Induction

$$\alpha_{t+1}(j) = [\sum_{i=1}^{N} \alpha_t(i) a_{ij}] b_j(\mathbf{x}_{t+1}), \quad 1 \leq t \leq T-1, \quad 1 \leq j \leq N. \tag{5.15}$$

3. Termination

$$P(\mathbf{X}|\lambda) = \sum_{i=1}^{N} \alpha_T(i). \tag{5.16}$$

1.3 HIDDEN MARKOV MODEL WITH CONTINUOUS OBSERVATION DENSITY

When observations are continuous signals/vectors, HMM with continuous observation densities should be used. In such a case, some restrictions must be placed on the form of the model probability density function (pdf) to ensure that pdf parameters can be updated in a consistent way. The most general pdf form is a finite mixture shown as follows:

$$b_j(\mathbf{x}) = \sum_{k=1}^{M} c_{jk} \mathcal{N}(\mathbf{x}, \mu_{jk}, \Sigma_{jk}), \quad 1 \leq j \leq N, \tag{5.17}$$

where $\mathbf{x}$ is the observation vector, c_{jk} is the mixture weight for the kth mixture in state j, and $\mathcal{N}$ is any log-concave or elliptically symmetric density. Without loss of generality, we assume that $\mathcal{N}$ is Gaussian with mean vector μ_{jk} and covariance matrix Σ_{jk} for the kth mixture component in state j. The mixture gains c_{jk} satisfy the stochastic constraint:

$$\sum_{k=1}^{M} c_{jk} = 1, \quad c_{jk} \geq 0, \quad 1 \leq j \leq N, \quad 1 \leq k \leq M. \tag{5.18}$$

By comparing (5.17) with the Gaussian mixture density given in (5.1), it is obvious that the Gaussian mixture model is actually one special case of the hidden Markov model with continuous observation densities, when there is only one state in the HMM ($N = 1$) and $\mathcal{N}$ is Gaussian. The parameter updating formulas for the mixture density, i.e. for c_{jk}, μ_{jk}, and Σ_{jk}, are the same as those for the GMM, i.e. formulas (5.5) - (5.8).

1.4 HIDDEN MARKOV MODEL WITH EXPLICIT STATE DURATION DENSITY

For many physical signals, it is preferable to explicitly model the state duration density in some analytic form. That is, a transition is made only after an appropriate number of observations occur in one state (as specified by the duration density). Such a model is sometimes called the semi-Markov model. We denote the possibility of d consecutive observations in state i as $p_i(d)$. Changes must be made to the formulas for calculating $P(\mathbf{X}|\lambda)$ and updating of model parameters. We assume that the first state begins at $t = 1$ and the last state ends at $t = T$. With the forward variable $\alpha_t(i)$ now defined as

$$\alpha_t(i) = P(\mathbf{x}_1, \mathbf{x}_2, \ldots, \mathbf{x}_t, \text{stay in state } i \text{ ends at } t|\lambda). \tag{5.19}$$

The induction steps for calculating $P(\mathbf{X}|\lambda)$ are given below.

1. Initialization

$$\alpha_1(i) = \pi_i p_i(1) \cdot b_i(\mathbf{x}_1), \quad 1 \leq i \leq N. \tag{5.20}$$

2. Induction

$$\alpha_t(i) = \pi_i p_i(t) \prod_{s=1}^{t} b_i(\mathbf{x}_s) + \sum_{d=1}^{t-1} \sum_{\substack{j=1 \\ j \neq i}}^{N} \alpha_{t-d}(j) a_{ji} p_i(d) \cdot \prod_{s=t+1-d}^{t} b_i(\mathbf{x}_s),$$

$$2 \leq t \leq D, \quad 1 \leq i \leq N, \tag{5.21}$$

and

$$\alpha_t(i) = \sum_{j=1}^{N}\sum_{d=1}^{D} \alpha_{t-d}(j)a_{ji}p_i(d) \cdot \prod_{s=t+1-d}^{t} b_i(\mathbf{x}_s),$$

$$D < t \leq T, \ \ 1 \leq i \leq N, \tag{5.22}$$

where D is the maximum duration within any state.

3. Termination

$$P(\mathbf{X}|\lambda) = \sum_{i=1}^{N} \alpha_T(i). \tag{5.23}$$

2. CLUSTERING OF FEATURE VECTORS

Before building the hidden Markov model, feature vectors in the training set of one sound class are clustered into several sets, with each set having distinct energy and spectral shape features from those of others. Each set will be modeled later by one state in the HMM. We chose an adaptive sample set construction method [55] and modified it to be used as the clustering tool. The resulting algorithm is stated as follows.

1. Define two thresholds: t_1 and t_2, with $t_1 > t_2$.

2. Take the sample with the largest norm (denote it as $\mathbf{x}_1$) as the representative of the first cluster: $\mathbf{z}_1 = \mathbf{x}_1$, where $\mathbf{z}_1$ is the center of the first cluster.

3. Take the next sample and compute its distance $d_j(\mathbf{x}, \mathbf{z}_j)$ to all the existing clusters (where $\mathbf{z}_j$ is the center of the jth cluster), and choose the minimum of d_j. Suppose $i = \arg\min\{d_j\}$.

   (a) If $\min\{d_i\} \leq t_2$, assign $\mathbf{x}$ to the ith cluster, and update the center of this cluster: $\mathbf{z}_i$.

   (b) If $\min\{d_i\} > t_1$, form a new cluster with $\mathbf{x}$ as the center.

   (c) If $t_2 < \min\{d_i\} \leq t_1$, do not assign $\mathbf{x}$ to any cluster, as it is in the intermediate region of clusters.

4. Repeat *Step 3* until all samples have been checked once. Calculate the variances of all the clusters.

5. If the variance is the same as last time, meaning the training process has converged, go to *Step 6*. Otherwise, return to *Step 3* for another iteration.

6. If there are still unassigned samples (in the intermediate regions), assign them to the nearest clusters. If the number of unassigned

samples is larger than a certain percentage of the total number of samples, adjust thresholds t_1 and t_2, and start with *Step 2* again.

The smoothed short-time power spectrum, which is the feature extracted from audio signal for denoting timbre as mentioned in Chapter 3, is separated into two parts. One part is the summation of the power spectrum over the frequency axis, which is the energy of the digital signal; the other part is the spectral envelope with unit energy, representing the frequency distribution of the signal. The clustering is first done according to the energy of each feature vector. Then, within each resulted cluster, a second clustering is conducted according to the spectral shape of feature vectors.

The number of states in the HMM can be adjusted by changing the threshold values in the clustering process. We tend to keep the number of states as such that different states should have distinct differences and physical meanings. And the slight differences within each state can be handled by the weights of the Gaussian mixtures. For example, setting $t_1 = 3$ and $t_2 = 2$, feature vectors in the training set for the sound of "dog bark" were first separated into four subsets based on the energy values, corresponding roughly to the sounds of bark, the intermissions, and the transition periods in between. Then, each subset was further divided into 2 - 4 states due to the spectral shape differences, with $t_1 = 10$ and $t_2 = 8$. As a whole, the HMM for the audio class of "dog bark" includes 12 states. And each feature vector in the training set is indexed with a state number.

For other examples, with the same set of thresholds as above, the training feature vectors of the sound class "applause" were first divided into 4 clusters by energy differences, and then each cluster was separated into 2 groups based on spectral shape, thus resulting in a total number of 8 states. And the training sets for the sound classes of "windstorm" and "rain" were clustered into 12 and 5 states, respectively.

3. TRAINING OF HMM PARAMETER SETS

3.1 THE TRAINING PROCESS

The training process is to estimate the HMM parameter set λ based on the training vector sequence $\mathbf{X}$, so that the likelihood $P(\mathbf{X}|\lambda)$ can get to a maximum. Widely used for HMM training is the Baum-Welch method [50], which is an expectation-maximization (EM) based iterative procedure and guarantees a monotonic improvement in the likelihood. However, the computational cost of this algorithm is very high. In order to achieve real-time processing and to leave enough space for further improvements of the model, we developed a simplified and much faster training process which includes four steps as described below.

First, we model the unit-energy spectral shape vector sequence in each state with the density depicted in formula (5.1). The vectors have a dimension of 65, representing the frequency distribution from 0 to π. The Gaussian mixture model parameters for the spectral envelope distribution are estimated for each state, respectively, at the first step of the training procedure. We select the number of mixtures in the GMM to be 5. The parameter sets

$$\beta_j = \{c_{jk}, \mu_{jk}, \Sigma_{jk}, \ 1 \leq k \leq M\}, \ \ 1 \leq j \leq N, \tag{5.24}$$

where N is the number of states, M is the number of mixtures, are estimated through the training algorithm of the GMM described in formulas (5.5) - (5.8), which is an iterative EM process.

There are two major motivations for using Gaussian mixture densities as the representation of audio features. The first one is the intuitive notion that the individual component of a multi-modal density such as GMM, may model some underlying set of acoustic classes. That is, the spectral shape of the ith acoustic class can in turn be represented by the mean $\vec{\mu}_i$ of the ith component density and variations of the average spectral shape can be represented by the covariance matrix Σ_i. The second motivation is the empirical observation that a linear combination of Gaussian functions is capable of representing a large class of sample distributions. Actually, one of the powerful attributes of GMM is its ability to form smooth approximations to arbitrarily-shaped densities.

In the second step, the distribution density for the energy value is estimated state by state. We choose the pdf form of energy to be the Gaussian density, i.e.

$$P_i(e) = \frac{1}{\sqrt{2\pi\sigma_i^2}} \exp\{-\frac{(e-\mu_i)^2}{2\sigma_i^2}\}, \ \ 1 \leq i \leq N, \tag{5.25}$$

where e is the energy value, and μ_i and σ_i^2 are estimated statistically from the energy values of the feature vectors in the ith state.

The third step is to calculate the state transition probability matrix $A = \{a_{ij}\}$ as such:

$$a_{ij} = t_{ij}/t_i, \ \ 1 \leq i, \ j \leq N, \tag{5.26}$$

where t_i is the number of transitions from state i to all other states, and t_{ij} is the number of transitions from state i to state j. The values of t_i and t_{ij} are computed from the state index of each feature vector in the training vector sequences.

Finally, since normally there is no restriction on which state the sound should start with, the initial state distribution is set as

$$\pi_i = 1/N, \quad 1 \leq i \leq N. \tag{5.27}$$

3.2 IMPLEMENTATIONAL ISSUES

Several implementational issues of the training process are discussed below.

1) Choosing Parameter Values. First, the number of components in the Gaussian mixture density M is normally determined by experiments. If M is too small, audio features may not be fully represented. On the other hand, if M is too large, it may cause problems in the training process, e.g. insufficient training data for a model of many parameters and an excessive amount of the computational cost. By observing the mean vectors and covariance matrices of component densities in each state with various selections of M, we found that $M = 5$ was the proper choice. Second, we selected diagonal covariance matrices for the component densities for the ease of computation. Note that full covariance matrices are not necessarily required in GMM, since the effect of using a set of full covariance Gaussians can be equally achieved by using a larger set of diagonal covariance Gaussians.

2) Initialization. We compared two ways of parameter initialization in the training of GMM. In the first approach, parameters were simply initialized to random values. That is, the initial mixture weights were random values between 0 and 1 satisfying $\sum_{i=1}^{M} p_i = 1$ for each state. Elements in the initial mean vectors were random values between 0 and -7, which corresponds to the normal range for element values in the spectral shape vectors. The initial diagonal elements in the covariance matrices are set to 1. In the second approach, the training feature vectors in each state were separated into M groups using the generalized Lloyd algorithm [56]. Then, the mean vector and the covariance matrix of each group are used as the initial mean and the initial covariance of one component in the Gaussian mixture, respectively. The ratio between the number of vectors in each group and the total number of vectors in the state is used as the initial mixture weight. It was found that the converged results from both initialization methods were very close to each other, while the second approach shortened the training process significantly. For example, in the training procedure of feature vectors in one particular state, it took 52 iterations to converge with the first type of initialization but only 8 iterations with the second type. It should be noted that an extra amount of computational cost was required by the generalized Lloyd iteration in the second approach.

3) Variance Limiting. When there are not enough data to sufficiently train a component's variance vector or when using noise-corrupted data, the variance elements can become very small which may produce singularities in computing the likelihood. To avoid such singularities, a variance limiting constraint is applied to the variance estimates after each EM iteration, which is as given below:

$$\overline{\sigma}_i^2 = \begin{cases} \sigma_i^2, & \text{if } \sigma_i^2 > \sigma_{\min}^2, \\ \sigma_{\min}^2, & \text{if } \sigma_i^2 \le \sigma_{\min}^2. \end{cases}$$

As the result of experiments, we choose $\sigma_{\min}^2 = 0.0001$ in our work.

4) Scaling. It is possible that the exponential item in (5.2) becomes very large, especially when the dimension of the feature vector is relatively high, and the Gaussian mixture density may become too small so that it exceeds the precision range of the computer. To keep numerical stability of the training process, a scaling factor $\exp\{c\}$ is calculated for each computation of the Gaussian mixture density, which is multiplied to every $b_i(\vec{x})$ in (5.1) to keep $p(\vec{x}|\lambda)$ from being too small. That is,

$$\begin{aligned} b_i'(\vec{x}) &= b_i(\vec{x}) \cdot \exp\{c\} \\ &= \frac{1}{(2\pi)^{D/2}|\Sigma_i|^{1/2}} \exp\{-(\frac{1}{2}(\vec{x}-\vec{\mu}_i)'\Sigma_i^{-1}(\vec{x}-\vec{\mu}_i) - c)\}, 1 \le i \le M, \end{aligned}$$

and

$$p'(\vec{x}|\lambda) = \sum_{i=1}^{M} p_i b_i'(\vec{x}) = \sum_{i=1}^{M} p_i b_i(\vec{x}) \exp\{c\}.$$

The parameter c is chosen large enough to ensure a sufficient precision of $p'(\vec{x}|\lambda)$. By using (5.8), we see that the scaling factor is canceled out in the *a posteriori* probability so that it does not affect the parameter updates. That is,

$$\begin{aligned} p'(i|\vec{x}_t, \lambda) &= \frac{p_i b_i'(\vec{x}_t)}{\Sigma_{k=1}^{M} p_k b_k'(\vec{x}_t)} = \frac{p_i b_i(\vec{x}_t)\exp\{c\}}{\Sigma_{k=1}^{M} p_k b_k(\vec{x}_t)\exp\{c\}} \\ &= \frac{p_i b_i(\vec{x}_t)}{\Sigma_{k=1}^{M} p_k b_k(\vec{x}_t)} = p(i|\vec{x}_t, \lambda). \end{aligned}$$

In computing the GMM likelihood $p(X|\lambda)$, we take the logarithm so that the term due to the scaling factor can be dealt with a subtraction:

$$\log\{p(X|\lambda)\} = \sum_{t=1}^{T} \log\{p(\vec{x}_t|\lambda)\} = \sum_{t=1}^{T} (\log\{p'(\vec{x}_t|\lambda)\} - c_t),$$

where $\exp\{c_t\}$ is the scaling factor for the feature vector $\vec{x}_t$.

3.3 COMPARISON WITH THE BAUM-WELCH METHOD

It should be noted that the above training method we developed is not a strict HMM process. In the original form of HMM, the information about which feature vector belongs to which state is hidden in both the training and the classification processes. In contrast, each feature vector is assigned to a state index according to the clustering result before the HMM parameter training in our approach.

By comparing the performance of our method with that of the Baum-Welch method, we observed that with the same initialization of parameter values, the convergent points of the two methods were very close to each other, and sometimes our method even had better results. However, the Baum-Welch method might require more computational time than our method up to dozens of times to achieve the same precision. This can be explained by the fact that our algorithm is performed based on the clustering result while the Baum-Welch method is based on the forward and backward procedures [50] which tend to be much slower. Therefore, the training procedure proposed here is more suitable for tasks in this research than the Baum-Welch method.

3.4 INCORPORATION OF THE VITERBI ALGORITHM

We also have done another experiment by designing an iteration between the training process and the Viterbi algorithm. The Viterbi algorithm is a formal technique to determine the single best state sequence, $\mathbf{q} = (q_1, q_2, \ldots, q_T)$, for a given observation sequence $\mathbf{X} = (\mathbf{x}_1, \mathbf{x}_2, \ldots, \mathbf{x}_T)$ and the HMM parameter set λ [50]. It includes an induction procedure which is similar in implementation to the forward calculation as described in formulas (5.13) - (5.16) as well as a backtracking step.

In our experiment, after the training of HMM parameters, a new state index for each feature vector in the training set is obtained by the Viterbi algorithm. Then, feature vectors are reclustered into states according to the new state indices. A new round of training for the HMM parameter set is then conducted based on these states as shown in Figure 5.1. Such iterations continue until a convergent point is reached. This procedure converges after a certain number of iterations (usually the more states in the HMM, the more iterations are needed). The convergent point is actually quite near to the maximum reached by using the training algorithm only once. But, improvements still can be observed. For example, the mean vectors tend to be more smooth and the mixture weights more evenly-distributed among component densities in results obtained by the iterated procedure than those from the original training process. However, the improvement was achieved at the cost of extra computation in the training iteration and the execution of the Viterbi algorithm.

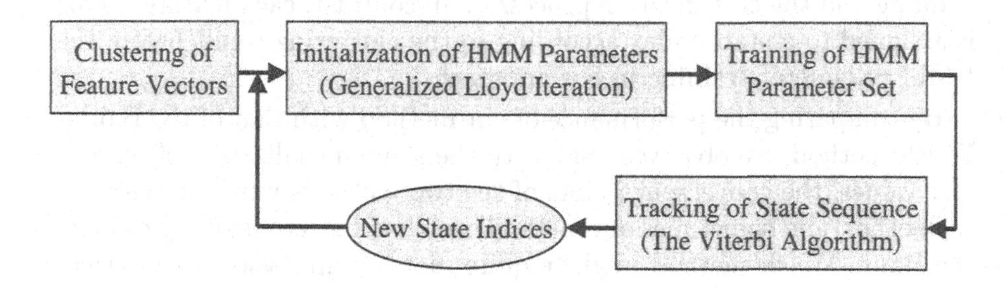

Figure 5.1. Iterative procedure for HMM parameter training with integration of the Viterbi algorithm.

4. CLASSIFICATION OF ENVIRONMENTAL SOUND

For the fine-level classification of environmental sound, we build a hidden Markov model for each semantic class of sound effect in the database. Assume that there are altogether K classes of sounds modeled with parameter sets λ_i, $1 \leq i \leq K$. For a piece of sound to be classified, feature vectors $\mathbf{X} = \{\mathbf{x}_1, \mathbf{x}_2, \ldots, \mathbf{x}_T\}$ are extracted from this sound. Then, the HMM likelihoods $P_i(\mathbf{X}|\lambda_i)$, $1 \leq i \leq K$, are computed. Choose the class j which maximizes P_i, i.e. $j = \arg\max\{P_i, 1 \leq i \leq K\}$, and the sound is classified into this class.

In computing $P(\mathbf{X}|\lambda)$, the forward procedure formulas (5.13) - (5.16) are used except that the observation probability $b_j(\mathbf{x})$ is computed as

$$b_j(\mathbf{x}) = \mathcal{N}(e, \mu_j, \sigma_j^2) \cdot \sum_{k=1}^{M} c_{jk} \mathcal{N}(\mathbf{s}, \mu_{jk}, \Sigma_{jk}), \quad 1 \leq j \leq N, \tag{5.28}$$

where $\mathbf{s}$ is the spectral shape vector and e is the energy value of the feature vector $\mathbf{x}$. That is, we define the observation probability of one feature vector $\mathbf{x}$ to be the product of the probability for the spectral shape and the probability for the energy of $\mathbf{x}$.

The problem of numerical stability also occurs in the implementation of classification. It can be seen from the forward procedure that as t becomes large, each term of $\alpha_t(i)$ starts to approach to zero exponentially. In order to keep a sufficient precision in the calculations of $\alpha_t(i)$

and $P(\mathbf{X}|\lambda)$, we apply scaling factors to terms of $b_j(\mathbf{x}_t)$, and take the logarithm of each term in the computation. Since there are addition operations in the original formulas, we have to use the exponential and logarithmic operations alternately and apply scaling factors before exponential operations. Therefore, the process is more complicated than that in the training procedure.

5. QUERY-BY-EXAMPLE RETRIEVAL OF ENVIRONMENTAL SOUND

In the query-by-example retrieval of sound effects, a hidden Markov model is built for each sound clip in the audio database. That is, feature vectors are extracted from one piece of sound, and are used to train the HMM parameter set for this sound. Then, with an input query sound, its feature vectors $\mathbf{X} = \{\mathbf{x}_1, \mathbf{x}_2, \ldots, \mathbf{x}_T\}$ are extracted, and the possibilities $P(\mathbf{X}|\lambda_i)$, $1 \leq i \leq L$, are computed according to the forward procedures, where λ_i denotes the HMM parameter set for the ith sound clip and L is the number of sound clips in the database. By comparing the values of $P(\mathbf{X}|\lambda_i)$, a ranking list of audio samples in the database in terms of similarity with the input query can be obtained.

IV

IMAGE SEQUENCE ANALYSIS

Chapter 6

IMAGE SEQUENCE ANALYSIS

1. HISTOGRAM DIFFERENCE VALUE IN IMAGE SEQUENCES

1.1 DEFINITION OF THE METRICS

In this research, we work on image sequences of the YUV format. The YUV coordinate system is adopted in the PAL and SECAM color television systems which are used in many countries. It consists of the luminance Y and two color components U and V as defined below [57]:

$$U = \frac{(B_N - Y)}{2.03} = 0.493(B_N - Y), \tag{6.1}$$

$$V = \frac{(R_N - Y)}{1.14} = 0.877(R_N - Y), \tag{6.2}$$

where B_N and R_N are intensities of the blue and red colors, respectively. Each of the three coordinates is represented with 8 bits so that there are 256 grey levels for each coordinate.

Let $H_i(k)$ and $H_j(k)$ denote the histogram values for the ith frame and the jth frame, respectively, and k is one of the 256 possible gray levels (for the Y-, U-, or V- component, or a combined code of the three components). Then, the difference between the ith and jth frames is often calculated by using the following two metrics:

$$D_1(H_i, H_j) = \frac{1}{256}\sum_{k=1}^{256} |H_i(k) - H_j(k)|, \tag{6.3}$$

or equivalently

$$D_2(H_i, H_j) = 1 - \frac{\sum_{k=1}^{256} \min(H_i(k), H_j(k))}{\sum_{k=1}^{256} H_i(k)}. \tag{6.4}$$

To reflect the difference between two frames more strongly, we can also try the following metric:

$$D_3(H_i, H_j) = \frac{1}{256} \sum_{k=1}^{256} (H_i(k) - H_j(k))^2. \tag{6.5}$$

It is observed in experiments that while metric D_3 enhances differences between abrupt shot changes and noises, the histogram difference values due to gradual shot transitions are also attenuated. For example, some dissolves and wipes are not revealed at all. Meanwhile, D_3 requires more computations than metrics D_1 and D_2.

1.2 HISTOGRAM DIFFERENCE OF THE Y-COMPONENT

The histogram differences between consecutive frames as computed with metric D_1 in a video clip containing 4000 frames are plotted in Figure 6.1 for the Y coordinate. This video clip is one part of a scientific program about bridge construction. All abrupt shot changes are reflected by remarkable pulses in this figure, which are marked with symbols A1, A2, ..., A10. There are also two gradual shot transitions (both are dissolves) marked by D1 and D2, where the shot does not change abruptly but over a period of a few frames. Several frames within the dissolve sequence of D2 are shown in Figure 6.2. It can be seen that the histogram difference values are quite small during the dissolve process, especially in the beginning frames, which makes it difficult to distinguish these shot changes from camera motions (such as pan, tilt, zoom-in, zoom-out), illumination changes, and various kinds of noises. For example, peaks to the left of D1 are caused by fast camera tilting and observed to be higher than the peak of D1.

The two peaks marked with L1 and L2 result from illumination changes during a camera panning motion. The two consecutive frames at L1, together with their histograms, are plotted in Figure 6.3. We can see that even though the two frames are quite alike, their histograms have a slight offset to each other due to the luminance shift. Actually, L1 has about the same height as that of A10 as shown in Figure 6.4. Although there is an abrupt change of contents between the two frames at A10, their histograms of the Y-component have similar distributions. We can hardly differentiate these two cases by observing the histogram difference values of the Y coordinate alone.

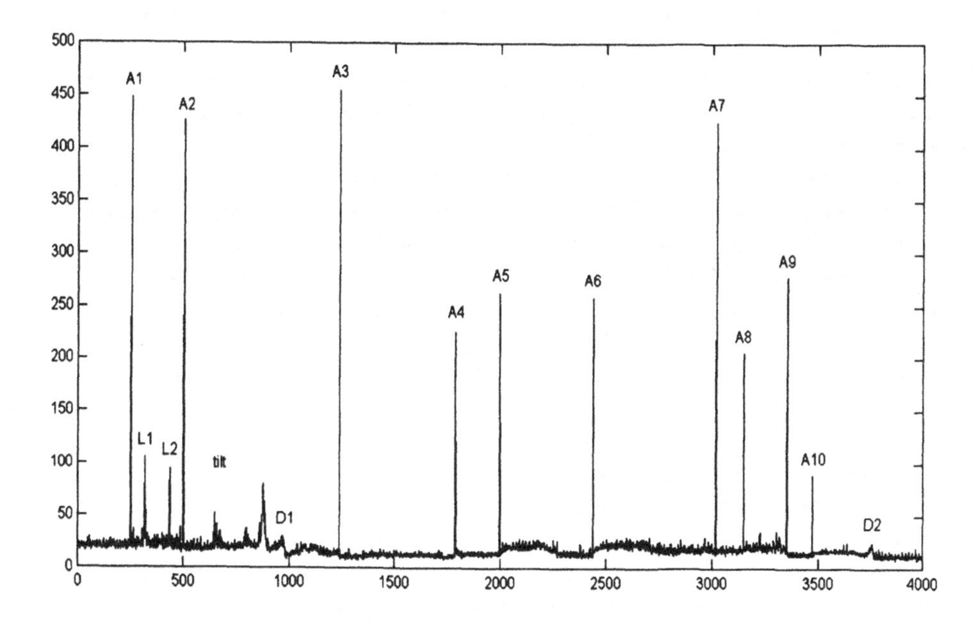

Figure 6.1. Histogram differences between consecutive frames for the Y-component in a video clip containing 4000 frames.

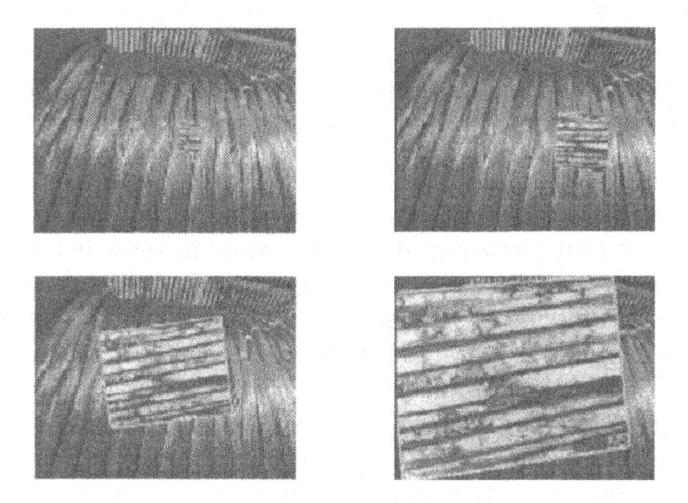

Figure 6.2. Frames within a gradual shot transition (dissolve) sequence.

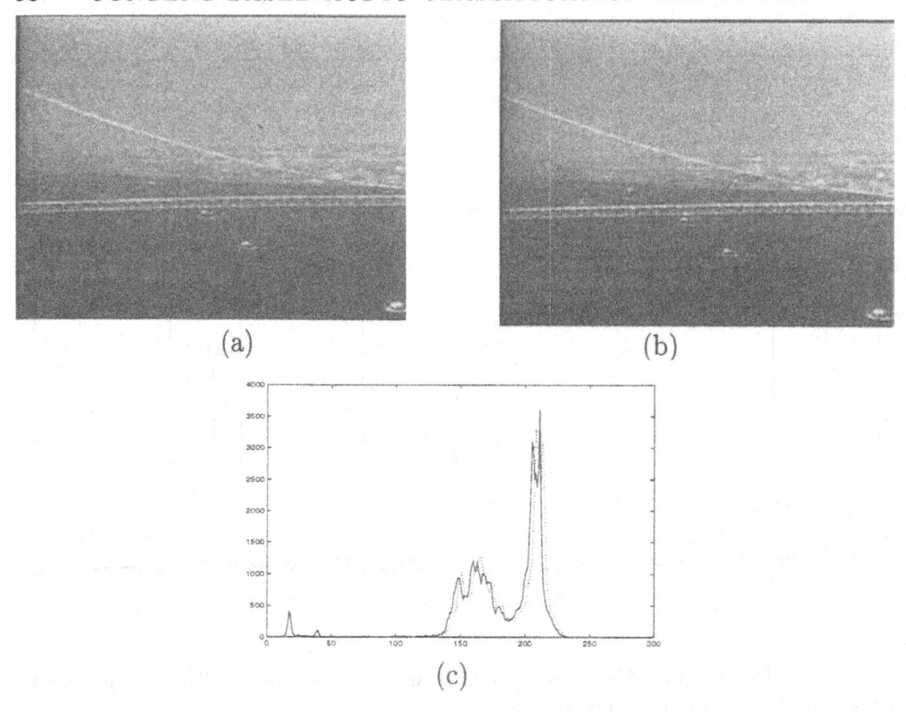

Figure 6.3. Histogram comparison between two consecutive frames in camera panning: (a) the former frame, (b) the latter frame, and (c) histograms of the two frames with a solid line for the former one and a dotted line for the latter one.

1.3 HISTOGRAM DIFFERENCE OF THE U AND V COMPONENTS

Histogram differences of the U- and V- components are plotted in Figures 6.5 and 6.6, respectively, for the same video clip as discussed above. We see that histogram difference values in these figures tend to be more noisy (i.e. with many pulses appearing within one shot) than those of the Y-component, especially during camera motions as indicated in the figures. However, abrupt shot changes are still prominent and easy to detect. Meanwhile, illumination changes are not as remarkable as in the Y-component. For example, the peak at L1 is much lower than the peak of A10 in the U and V histogram difference values. What is even more important is that gradual transitions which are not obvious in the Y-component may be revealed more clearly in the U- and V-components. As shown in these figures, peaks at D1 and D2 are much higher than those of neighboring values, especially in the V-component histogram difference. Therefore, information revealed in the U and V coordinates can compensate that in the Y coordinate in differentiating many cases.

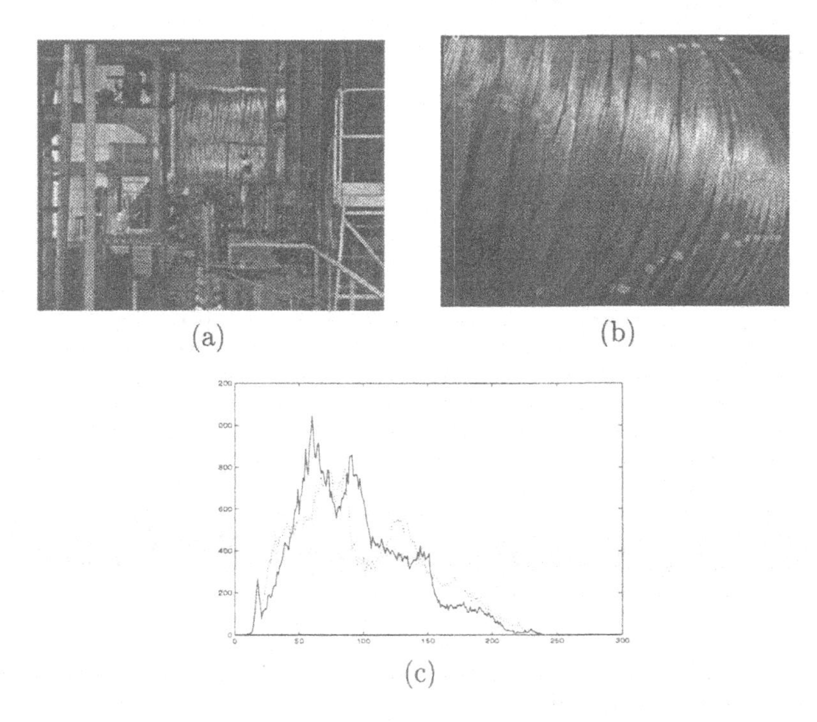

Figure 6.4. (a) The frame before the shot cut, (b) the frame after the shot cut, and (c) histograms of the two frames with a solid line for the former one and a dotted line for the latter one.

1.4 HISTOGRAM DIFFERENCE OF THE COMBINED CODE

As an effort to integrate the three coordinates, a combined code of 8 bits is built by taking the most significant 4 bits from the Y-component, and 2 bits from each of the U- and V- components. Histogram differences of the combined code are shown in Figure 6.7. Compared to those of the Y-component, the histogram differences of the combined code are quite noisy; while compared to those of the U- and V- components, some gradual transitions may be less obvious in these values (note the peak of D1 in this figure). Consequently, it is even more difficult to select proper thresholds for detecting gradual shot changes with the combined code than computing histogram difference for each of the three coordinates separately.

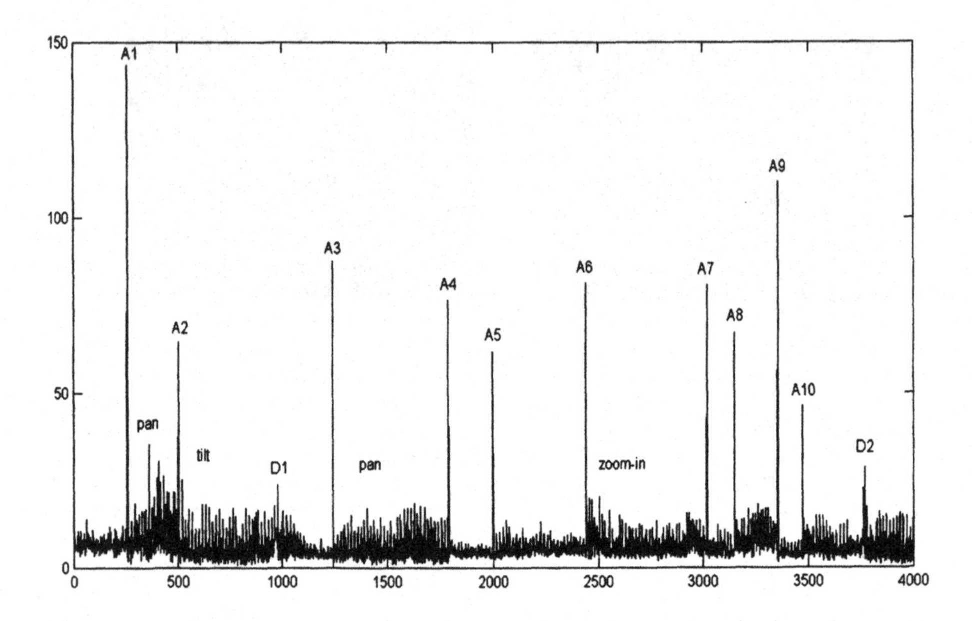

Figure 6.5. Histogram differences between consecutive frames for the U-component in a video clip containing 4000 frames.

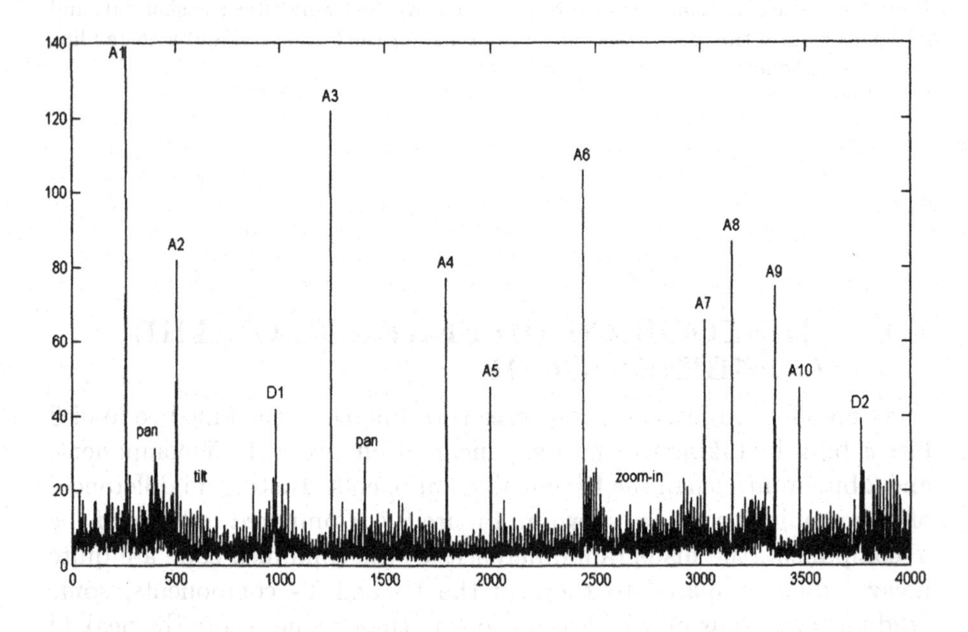

Figure 6.6. Histogram differences between consecutive frames for the V-component in a video clip containing 4000 frames.

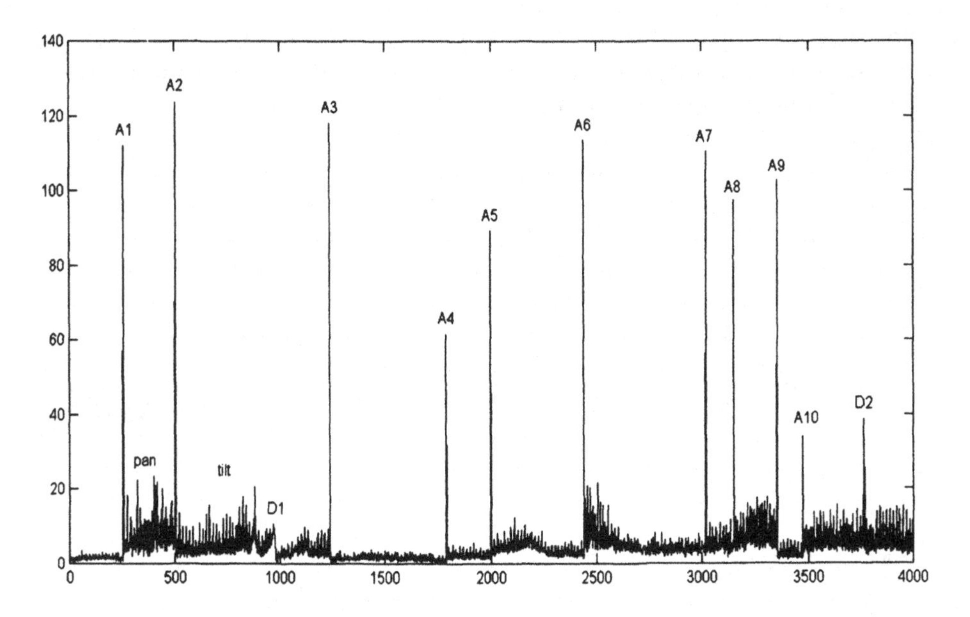

Figure 6.7. Histogram differences between consecutive frames for the combined code in a video clip containing 4000 frames.

2. THE TWIN-COMPARISON APPROACH

2.1 THE ORIGINAL ALGORITHM

In order to detect both abrupt and gradual shot changes, Zhang *et al.* proposed a twin-comparison approach which takes into account cumulative differences between frames for gradual transitions [6]. This method requires the use of two cutoff thresholds: a higher threshold T_h for camera break detection and a lower threshold T_l for special effect detection. The process begins by comparing consecutive frames with a difference metric such as (6.3). Whenever the difference value exceeds threshold T_h, a camera break is declared. However, the twin-comparison also detects differences which are smaller than T_h but larger than T_l. Any frame that has such a difference value with regard to its previous frame is marked as the potential start (F_s) of a gradual transition. It is then compared to subsequent frames as shown in Figure 6.8. This accumulated difference will normally increase during a gradual transition. The end frame (F_e) of the transition is detected when the accumulated difference increases to a value larger than T_h and the difference between consecutive frames decreases to less than T_l. If the difference between consecutive frames drops below T_l before the cumulative difference exceeds T_h, then the potential frame is dropped and the search starts all over. There are some gradual transitions in which the difference falls below the lower threshold. The user can set a tolerance value to allow a certain number of consecutive frames to fall below the threshold.

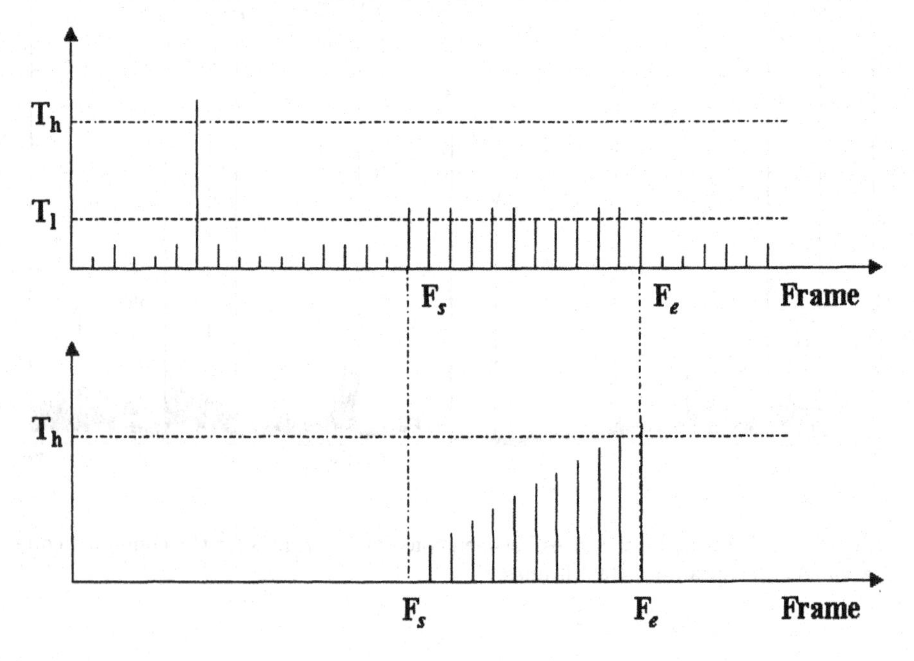

Figure 6.8. Illustration of the twin-comparison method for the detection of gradual shot transitions.

A key issue in applying the twin-comparison algorithm is to select appropriate threshold values. In [6], the selection of T_h and T_l is based on the normalized frame-to-frame differences over an entire given video source. Let σ be the standard deviation and μ the mean of the frame-to-frame differences. The threshold T_h is selected as

$$T_h = \mu + \alpha\sigma, \tag{6.6}$$

where α is empirically chosen between five and six. The lower threshold T_l should be generally larger than the mean value of the frame-to-frame differences of the entire video package. It does not vary significantly based on the examination of three documentary videos.

2.2 EXPERIMENTAL RESULTS AND MODIFICATIONS

We conducted experiments of the twin-comparison method on five video clips consisting of news bulletin, cartoon, documentary, and feature movie. Each video clip contains 6000 frames, and there are various kinds of shot transitions, camera motions, luminance changes and noise within these frame sequences. Histogram differences of the Y-component are computed only. The following observations and modifications are made based on experimental results.

The first issue is the selection of thresholds. It is found that the experience for choosing thresholds as introduced above only fits for documentaries with little noise. For the five test video clips, the value of α to estimate T_h was in the range of three to six, while the optimal lower threshold T_l varied from 20 to 80. Furthermore, to select thresholds with (6.6), the whole video sequence should be known *a priori*, which is not applicable for on-line processing of video content. Actually, the mean and variance of frame-to-frame differences may change significantly during one video program. Shown in Figures 6.9 and 6.10 are consecutive histogram differences for a cartoon video clip and a news video clip, respectively. Note that the range of values in the cartoon clip changes over time due to the complex content transitions and noise (indicated with the following symbols A: abrupt change, D: dissolve, L: luminance change, FO: fade-out). While the figure of the news video clip is simpler, the heights of peaks at shot changes vary dramatically, and there is strong noise during several shots. It is rather difficult to select thresholds suitable for the entire video clip. To overcome this difficulty, we propose the use of adaptive thresholds which are updated with new coming data. A sliding window containing 1000 frames is used that proceeds at an interval of 500 frames. Statistics of the histogram difference values as well as the amplitudes of pulses are integrated to determine thresholds. More discussions on selecting thresholds by combining Y- and V- components will be presented in the next section.

The second problem is the estimation of camera motions. Since changes resulting from camera movements tend to induce successive difference values of the same order as those of gradual shot transitions, the use of the twin-comparison algorithm alone is not enough to differentiate these two cases. For example, in one documentary video clip, there are fifteen shot changes including both abrupt and gradual transitions. Fourteen of them were detected with the twin-comparison approach and one was missed. Meanwhile, there were seven false alarms, among which five were resulted from camera tilting, one from panning, and one from zooming-in. In [6], the feature of optical flow was used to detect camera movements, which involves representing the difference between two consecutive frames as a set of motion vectors. During camera panning or tilting, these vectors will predominantly have the same direction. In the case of zooming, vertical components of motion vectors for the top and bottom rows of a frame should have opposite signs. However, to de-

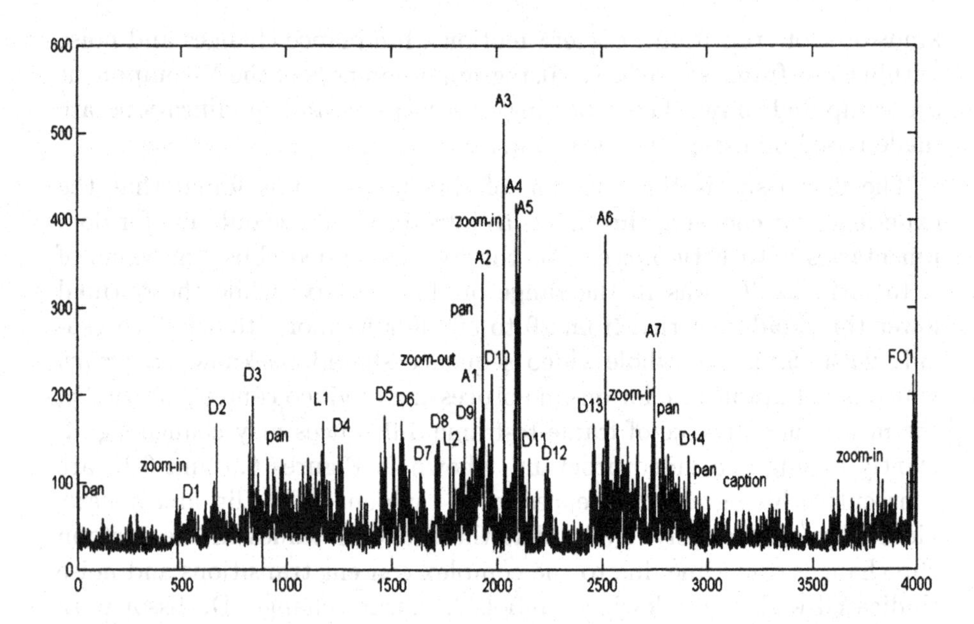

Figure 6.9. **Histogram differences between consecutive frames for the Y-component in a cartoon video clip.**

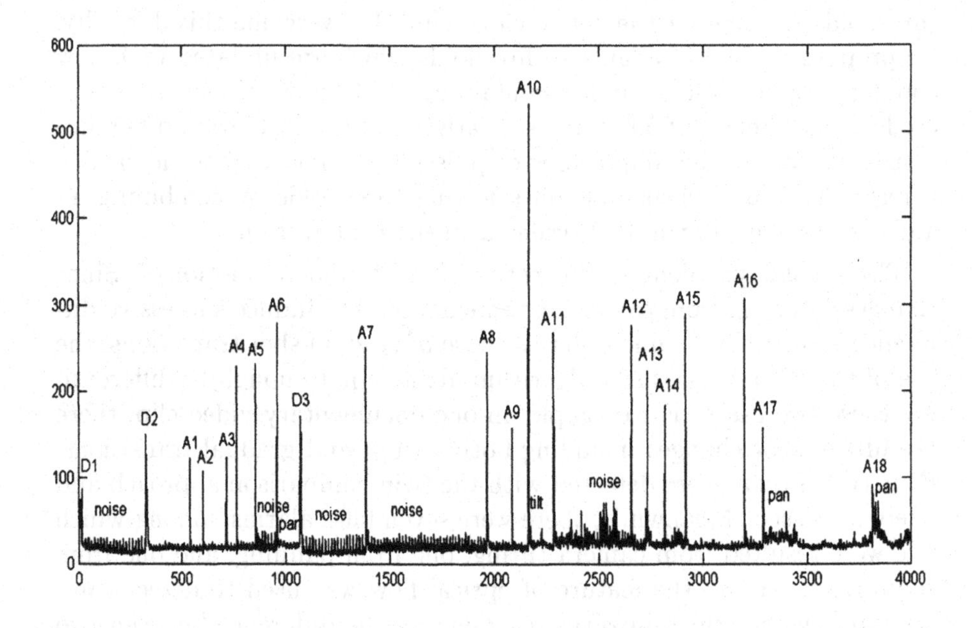

Figure 6.10. **Histogram differences between consecutive frames for the Y-component in a news video clip.**

termine a set of motion vectors for a frame is a time consuming process and may not be applicable in real-time processing.

Here, we propose a simplified and efficient method to detect camera motions. For camera panning and tilting, only the movement of a 50×50 block in the center of a frame is checked. During such movements, the central part of a frame will remain in the next frame with a shift (in the horizontal direction in panning and in the vertical direction in tilting). Correlation between the central blocks of two consecutive frames is calculated as

$$C_i(m,n) = \sum_{j}^{50} \sum_{k}^{50} |p_i(j,k) - p_{i+1}(j+m, k+n)|,$$

$$-20 \leq m \leq 20, \; -20 \leq n \leq 20, \tag{6.7}$$

where $p_i(j,k)$ is the pixel intensity at coordinates (j,k) in the central block of the ith frame. The coordinates (m_i, n_i) satisfying

$$(m_i, n_i) = \arg\min\{C_i(m,n), -20 \leq m \leq 20, -20 \leq n \leq 20\}, \tag{6.8}$$

give an estimation of the motion vector for the ith frame. The estimated horizontal shifts and correlation values between consecutive frames during a camera panning motion and the estimated vertical shifts and correlation values during a camera tilting motion are plotted in Figures 6.11 and 6.12, respectively.

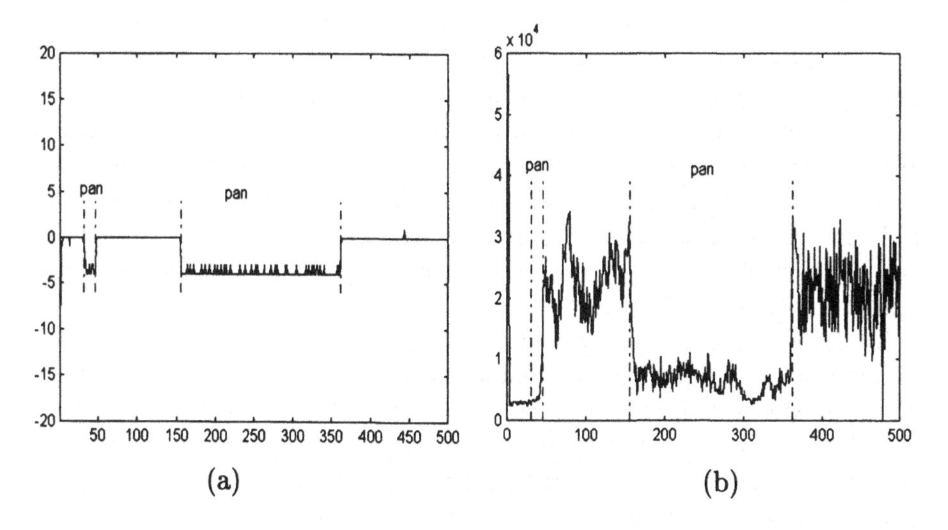

Figure 6.11. Detection of camera panning: (a) the horizontal shift of the central block and (b) the correlation value of the central block.

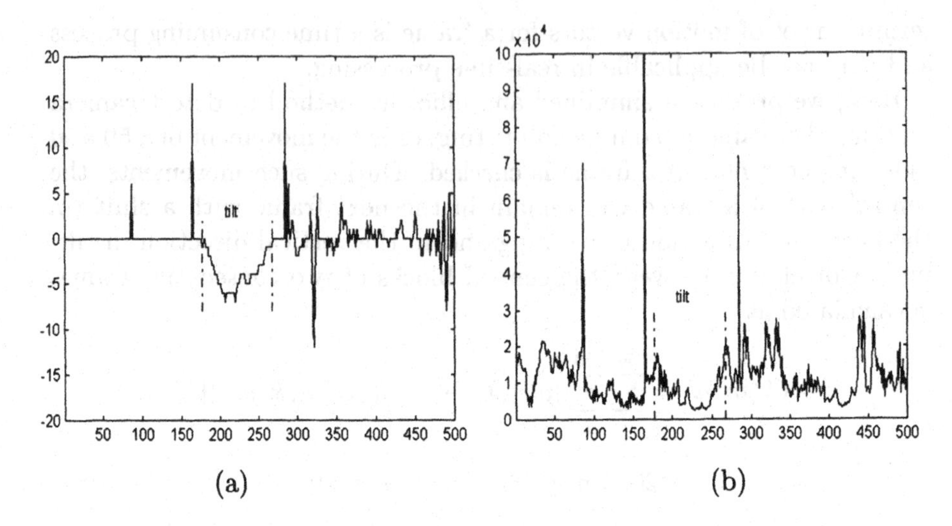

Figure 6.12. Detection of camera tilting: (a) the vertical shift of the central block and (b) the correlation value of the central block.

It is observed that during panning or tilting movements, the horizontal and vertical shifts change gradually or remain at a certain level depending on the speed and angle of camera motions. Meanwhile, the correlation values are kept at low levels during these motions because a rather good match can be found between two consecutive frames for the central block (while the correlation can be much higher during shot changes, object movements within one shot, or with the existence of noise). Therefore, by checking the values of correlation and shifts, a label can be assigned to each frame indicating whether it is within a camera motion of panning or tilting. Actually, it is not necessary to examine every frame in one video clip. We may apply the twin-comparison approach first, and take the detected gradual shot transitions as potential shot changes. Then, several frames around each potential transition can be checked to determine whether it is really a shot transition or just camera motion within one shot.

To detect zooming-in and zooming-out, we select two 20×20 blocks at the upper-middle and lower-middle parts of one frame, respectively. Then, motion vectors of these blocks are estimated by calculating correlation values between the current frame and the next one in a range of -20 to 20 in both the horizontal and vertical directions for each block. During camera zooming motions, vertical shifts of the two blocks should have about the same amplitude but different signs. Thus, a frame can be labeled as in a zooming motion or not by checking vertical shifts of upper and lower blocks. Again, the check of zooming is only done on frames around potential gradual transitions detected with the twin-comparison method. In this way, identification of camera motions just takes a little computation in the whole shot change detection process.

Performances of the modified twin-comparison approach on five test video clips are shown in Table 6.1. The sensitivity rate and the recall rate of detection are defined as

$$\text{sensitivity} = \frac{\text{No. of correct detects}}{\text{No. of correct detects} + \text{No. of missed detects}} \times 100\%, \tag{6.9}$$

$$\text{recall} = \frac{\text{No. of correct detects}}{\text{No. of correct detects} + \text{No. of false alarms}} \times 100\%. \tag{6.10}$$

Table 6.1. Performance of the modified twin-comparison approach.

Testing	Actual Number	Correct detects		False alarms	
video clips	of shot changes	Number	Sensitivity	Number	Recall
News video	30	29	97%	3	91%
Cultural video	55	47	85%	5	90%
Movie	20	19	95%	1	95%
Scientific video	15	13	87%	1	93%
Cartoon	47	46	98%	10	82%
Total	167	154	92%	20	89%

Among the 13 shot changes missed in the detection, ten were gradual transitions (including eight dissolves, one wipe and one fade out). Most were followed immediately by camera motions (or even occurring at the same time), which made the situation difficult to handle. The other three missed shot transitions were abrupt changes but with histogram difference value lower than the higher threshold. Most of false alarms were caused by illumination changes, appearance/disappearance of caption, and quick object movements.

3. SHOT CHANGE DETECTION BASED ON COMBINED Y- AND V- COMPONENTS

3.1 DETERMINATION OF THE LOWER THRESHOLD

As discussed earlier, histogram difference values can be quite small during many gradual shot transition procedures. In order to detect these shot changes, the lower threshold in the twin-comparison method should be selected to be a very small value, which turns out to be lower than many histogram differences within one shot. This implies that there will be a lot of computations wasted on the calculation of cumulative histogram differences between frames where there are actually no shot changes. Furthermore, some gradual transitions are very close to, or even mixed with, camera motions, which makes it difficult to detect these shot changes by analyzing histogram differences of the Y-component alone.

Nevertheless, it is observed that there are often high pulses reflecting the gradual shot changes in the histogram difference of the V-component. For example, among the ten missed gradual transitions in the above experiment, seven are with prominent peaks in the V-component histogram difference which can be easily detected with one threshold. Thus, we propose to compute histogram difference for Y- and V- components separately, and combine information in the two components to detect shot changes. High pulses occurring in the V-component difference but not in the corresponding Y-component difference will trigger a second check in the Y histogram differences via cumulative difference calculation. Although the calculation of V-component histograms requires an additional cost, the lower threshold T_l can be raised to a higher value so that a less amount of computation will be spent on determining cumulative differences. Thus, the overall complexity may still be low.

3.2 DETERMINATION OF THE HIGHER THRESHOLD

Although almost all abrupt shot changes are reflected by remarkable pulses in the Y-component histogram difference, some of them are not high enough to be distinguished from peaks produced by dramatic illumination changes, appearance/disappearance of caption, or fast object motions. This makes it difficult to select the higher threshold T_h, i.e. a hard trade-off must be made between the sensitivity and recall rates. To deal with such cases, the information revealed by the histogram differences of the V-component is again valuable, because many abrupt shot changes with shorter peaks in the Y-component difference may be much more prominent in the V-component difference. For instance, all the three abrupt changes that were missed in the above experiment are revealed with high pulses in the histogram difference of the V-component and can be easily detected. As the two consecutive frames before and after any abrupt shot change should have very different content, if it is not obviously reflected in the Y-component, then very likely it will be more apparently revealed in the V-component. Thus, by combining histogram differences of more than one component, we may choose the higher threshold for the Y-component difference to be slightly larger to avoid false alarms, and further check can be done on positions with remarkable pulses in the V-component difference but not having an amplitude higher than the threshold in the Y-component difference.

3.3 FRAMEWORK OF THE PROPOSED SCHEME

The proposed scheme for shot change detection is illustrated in Figure 6.13. Re-checking of the Y-component histogram is done in two ways. When the histogram difference value is below the lower threshold T_l, cumulative differences will be computed to check whether there is a missed gradual shot transition. When the difference value is higher than T_l but lower than T_h, histograms of the two consecutive frames before and after the peak will be examined to see whether it is caused by a real shot change, or results from illumination change, appearance/disappearance of caption, or object movement within one shot.

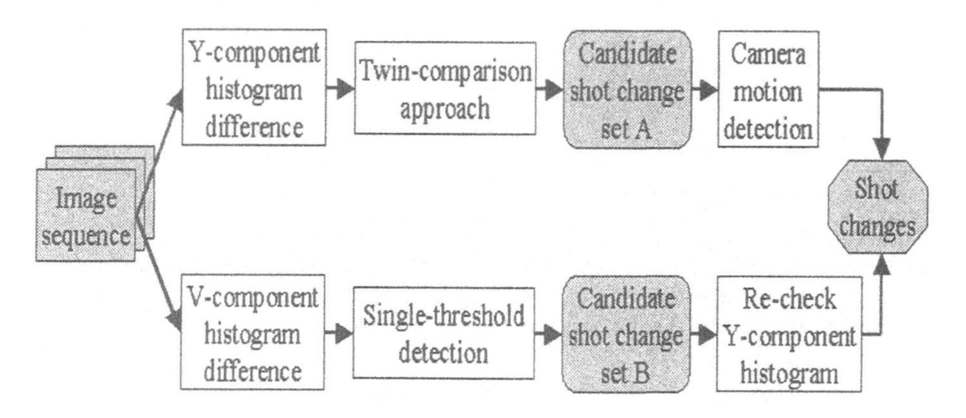

Figure 6.13. The framework of the proposed scheme for shot change detection.

Illumination change is one major source of false alarms in shot change detection. The histogram difference of the Y-component is especially sensitive to this kind of change. For example, among the ten false alarms in the shot change detection of the cartoon video, nine were originated from illumination changes which occurred frequently in this video clip. However, they were not so obviously reflected in the histogram difference of the V- or U- component. Actually, seven of them could not be revealed at all in the V-component difference. One typical instance is shown in Figure 6.14. The regular pattern for histograms of two consecutive frames with illumination change is that they are almost the same at the lower intensities while having a shift to each other at the higher intensities. Thus, in re-checking the Y-component histogram, this regularity is used to detect the luminance change.

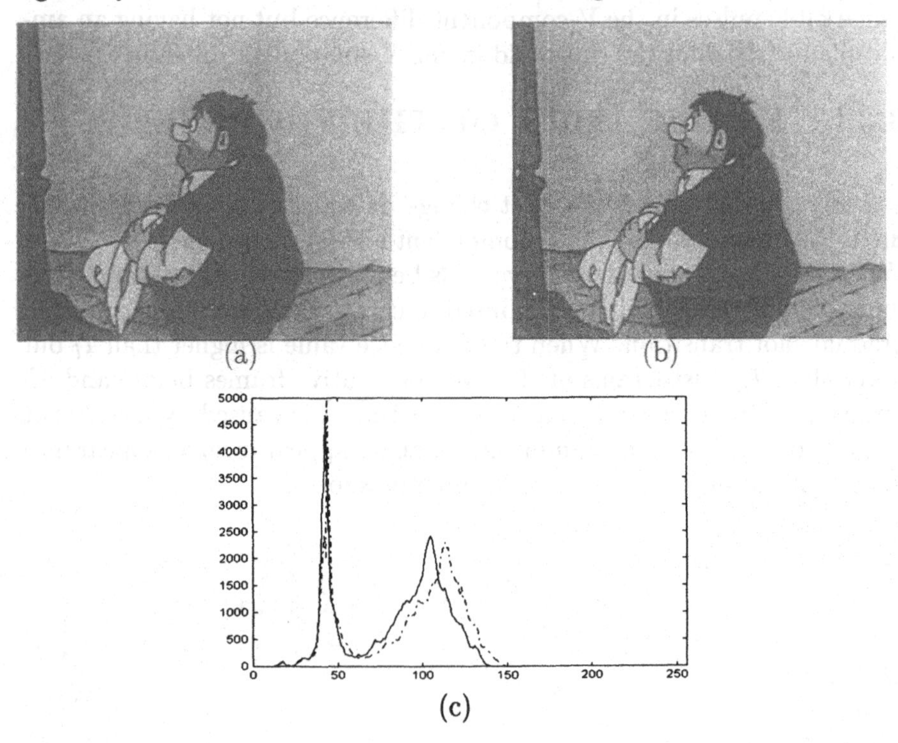

Figure 6.14. Illumination change between two consecutive frames: (a) the former frame, (b) the latter frame, and (c) histograms of the two frames with the solid line for the former one and the dashdot line for the latter one.

One special case is shown in Figure 6.15, where there is strong luminance change due to photoflash. Such a situation often appears in news bulletins. It can be seen that although the frame with photoflash has a quite different histogram to those of its neighboring frames, the histograms of its previous frame and its next one are very similar. Therefore, we can use this fact to distinguish illumination change caused by photoflash.

When the caption appears or disappears as shown in Figure 6.16, the value of histogram difference between the two consecutive frames, one with caption and the other without, is normally lower than the higher threshold. However, sometimes it may be at about the same level of the higher threshold or even higher. Nevertheless, the difference is concentrated in a narrow region of the histogram depending on the color of the caption, (e.g. between 200-250 when the caption is in white), while most parts of the histograms are similar. This fact can be used

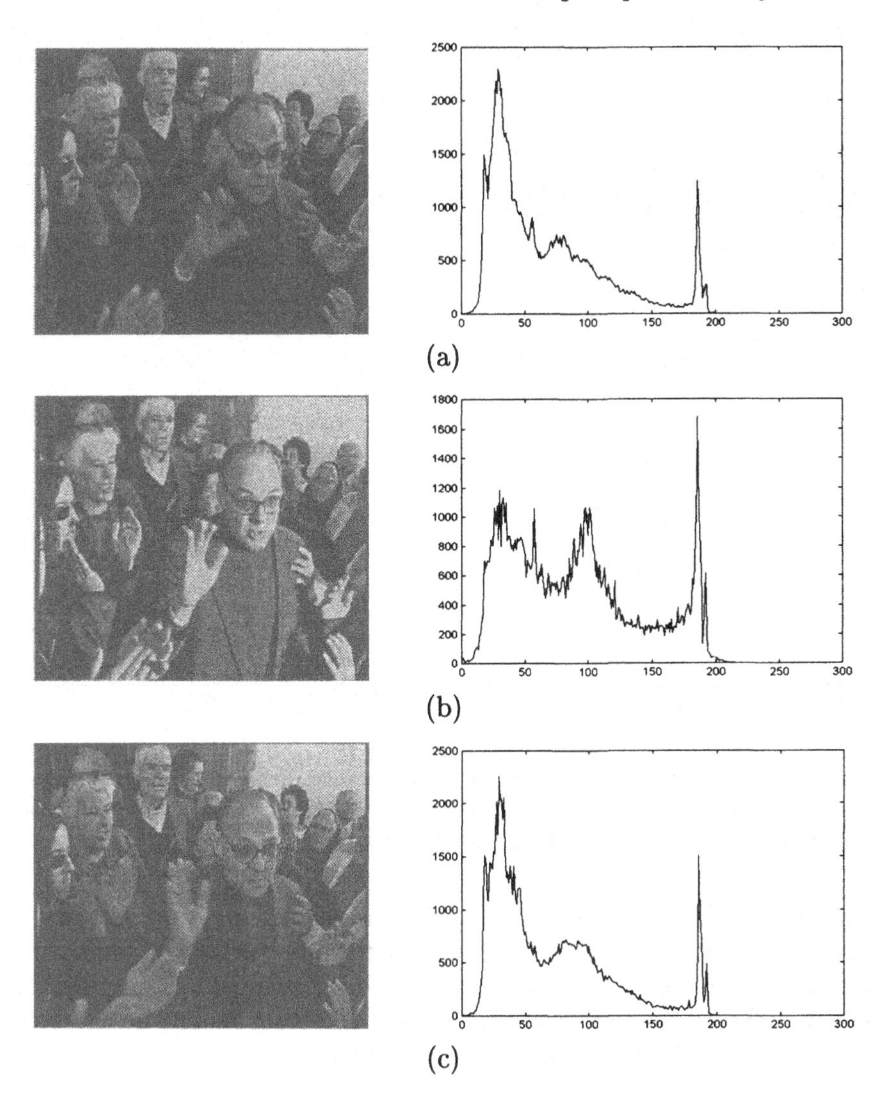

Figure 6.15. Illumination change due to photoflash: (a) the frame before photo-flash, (b) the frame with photoflash, and (c) the frame after photoflash.

to differentiate this case. Moreover, as caption normally appears only at the beginning or ending parts of a movie or video program, we may choose to detect the appearance/disappearance of caption only in certain temporal intervals of a video package.

Another reason for false positive detection is object movement (of higher speed or larger size) within one shot. A robust approach was presented in [58] which can be applied to such cases. It is based on the assumption that the movement influences no more than half of an

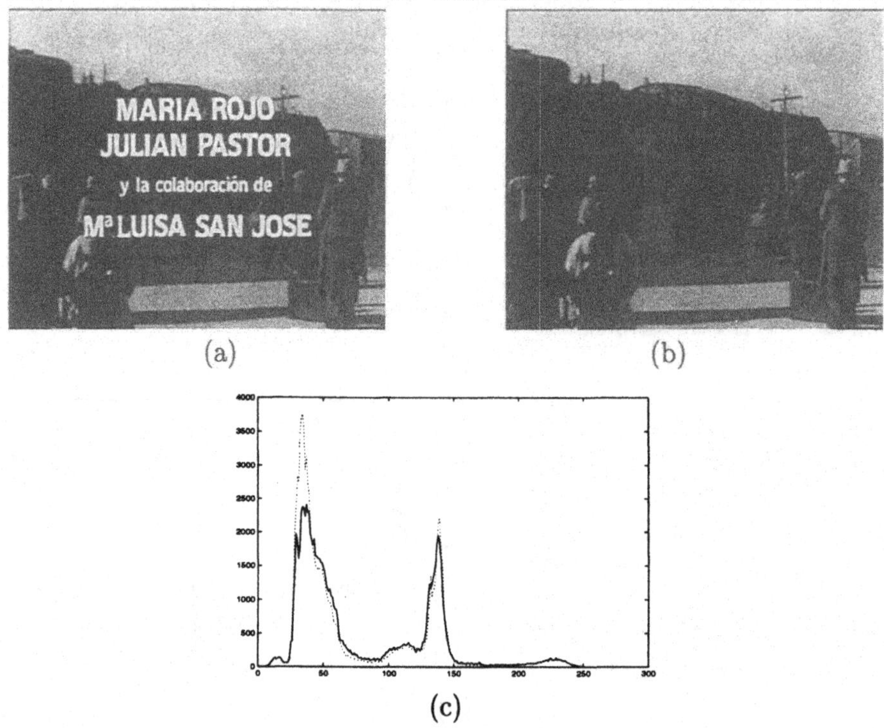

Figure 6.16. Histogram change due to disappearance of caption: (a) the former frame, (b) the latter frame, and (c) histograms of the two frames with the solid line for the former one and the dotted line for the latter one.

entire frame. When re-checking the Y-component histogram, a frame is divided into a 4×4 grid of 16 rectangular regions. Instead of comparing entire frames, histograms of the corresponding regions are compared to yield 16 difference values. The shot change detection is done by checking the sum of only the eight lowest difference values.

4. ADAPTIVE KEYFRAME EXTRACTION AND ASSOCIATED FEATURE ANALYSIS

4.1 ADAPTIVE KEYFRAME EXTRACTION

A simple and common way for keyframe extraction is to choose the first and/or the last frame of each shot. Actually, situations are quite diverse among different shots. Some shots may only contain frames which are very similar to each other while others may have large-scope camera motions in which the frame content changes dramatically. To handle various situations, we propose to use an adaptive scheme for keyframe extraction. That is, the first frame of one shot is chosen as the first keyframe. Then, its histogram is used as the reference to compare with those of latter frames. When the difference is higher than a predefined threshold, a second keyframe is selected. Next, the second keyframe will be used as the reference and compared with latter frames. This procedure continues until the end of the shot. In this way, one shot may have only one keyframe or have multiple keyframes depending on the complexity involved in it.

We may also define a maximum number of keyframes within each shot. Suppose that there are N frames extracted in the above threshold-based process, and the maximum number of keyframes allowed for one shot is K. When $N > K$, these N frames are taken as candidate keyframes. The first frame and the Nth frame are selected as the first keyframe and the Kth keyframe, respectively. Then, the $\frac{K+1}{2}$th keyframe (e.g. the 3rd keyframe when $K = 5$) is searched among frames between the first frame and the Nth frame. The one to be chosen should have approximately the same difference from the first frame as from the Nth frame as illustrated in Figure 6.17. Assume that it is the mth frame with $1 < m < N$. Next, the $\frac{K+3}{4}$th keyframe and the $\frac{3K+1}{4}$th keyframe (i.e. the 2nd keyframe and the 4th keyframe when $K = 5$) are searched in the same way within the region from the first to the mth frame, and that from the mth to the Nth frame, respectively. This procedure continues until all K keyframes are found.

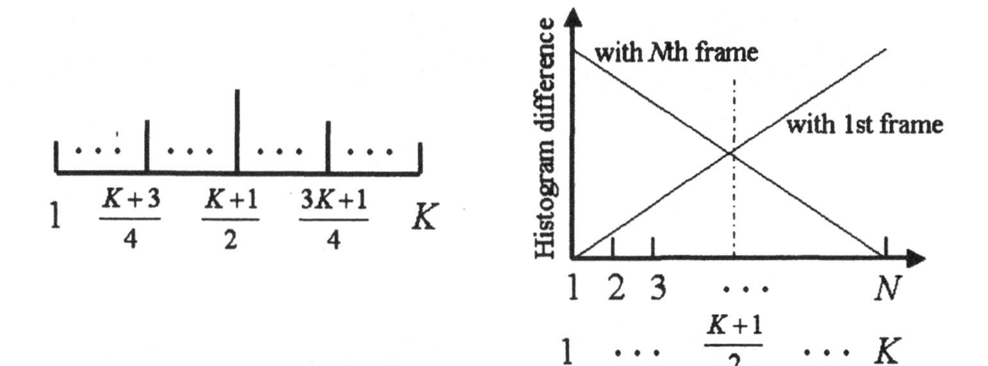

Figure 6.17. Selecting K final keyframes from N candidate keyframes ($K < N$).

4.2 FEATURE ANALYSIS OF KEYFRAMES

Many features have been investigated for the representation of the image or video content such as color, texture, shape and motion. We select only color histograms of keyframes for typical shots as the first step (e.g. anchorperson shots in news bulletin, shots of football or basketball game in sports video) to be features included in related video models in this work. Keyframes from corresponding video types will be matched with these histogram patterns to decide whether they belong to typical shots or not.

V

EXPERIMENTAL RESULTS

Chapter 7

EXPERIMENTAL RESULTS

1. GENERIC AUDIO DATA SEGMENTATION AND INDEXING

1.1 AUDIO DATABASE

We have built a generic audio database to be used as the testbed of the proposed algorithms, which consists of the following contents: 1000 clips of environmental audio including the sounds of applause, animal, footstep, raining, explosion, knocking, vehicles and so on; 100 pieces of classical music played with 10 kinds of instruments, 100 other music pieces of different styles (classic, jazz, blues, light music, Chinese and Indian folk music, etc.); 50 clips of songs sung by male, female, or children, with or without musical instrument accompaniment; 200 speech pieces in different languages (English, German, French, Spanish, Japanese, Chinese, etc.) and with different levels of noise; 50 clips of speech with the music background; 40 clips of environmental sound with the music background; and 20 samples of silence segment with different types of low-volume noise (clicks, brown noise, pink noise and white noise). These short pieces of sound clips (with duration from several seconds to more than one minute) are used to test the audio classification performances. We also collected dozens of longer audio clips recorded from movies or video programs. These pieces last from several minutes to half an hour, and contain various types of audio. They are used to test the performances for audiovisual data segmentation and indexing.

1.2 COARSE-LEVEL CLASSIFICATION RESULTS

The proposed classification approach for generic audio data achieved an accuracy rate of more than 90% by using a set of 1200 audio pieces including all types of sound selected from the audio database described above. Listed in Tables 7.1 and 7.2 are results at the two classification hierarchies, respectively, where sensitivity rate is defined as the ratio between the number of correctly classified samples and the actual number of samples in one category, and recall rate is the ratio between the number of correctly indexed samples and the total number of samples as indexed in one audio type (including false alarms). "MBG" is the abbreviation for "music background", and "SFX" is for "sound effects". In order to obtain threshold values which are used in the heuristic procedures, 10-50% of samples in each audio type were randomly selected to form a training set (i.e. there were 20-60 samples from each type). The threshold values were determined step by step according to the segmentation and indexing process as outlined in Section 5. And in each step, an iterative procedure of modifying and testing the threshold values was conducted until an optimal result was achieved. Then, the whole data set was used to testify the classification performances.

Table 7.1. Classification results for audio categories.

Audio Category	Test Samples	Correct Samples		False Alarms	
	Number	Number	Sensitivity	Number	Recall
Silence	20	20	100%	0	100%
With music	380	362	95.3%	0	100%
Without music	800	800	100%	18	97.8%

From Table 7.1 we can see that sounds without music components are correctly categorized since they all have a rather high zero ratio. For sounds in the first category, there are classification errors with some song segments (especially those without instrument accompaniment) and some speech with music background segments in which the music components are weak. However, with hybrid-type sounds and sound effects involved here, the overall accuracy for categorizing music/non-music sounds is still comparable to previous results for pure speech/music discrimination tasks.

Within the first category, several harmonic sound effects are indexed as "pure music", and there are music segments misclassified as song, speech with MBG or sound effect with MBG. A couple of song segments which lack the ripple-shaped spectral peak tracks are taken as sound effect with MBG. For the second category, apart from false alarms from the first category, there are also 27 misclassifications between pure speech and non-harmonic environmental sound segments. It should be noted that most mistakes result from the very noisy background in some speech, music and song segments. While our approach is normally robust in distinguishing speech and music with a rather high level of noise, the algorithm needs to be further improved so that speech and music components are correctly detected as long as their contents can be recognized by human perception. On the whole, the ratio of the total number of correctly indexed samples in the eight audio types (i.e. audio types listed in Table 7.2 plus silence) to the size of the data set reaches 94.8%. We also calculated the accuracy rate of samples not included in the training set, which is 90.7%. A demonstration program was made for the on-line audio classification, which shows the waveform, the audio features and the classification result for a given sound, as illustrated in Figure 7.1.

Table 7.2. Classification results for basic audio types.

Audio Type	Test Samples	Correct Samples		False Alarms	
	Number	Number	Sensitivity	Number	Recall
Pure speech	200	182	91%	16	91.9%
Pure music	200	189	94.5%	4	97.9%
Song	50	42	84%	2	95.5%
Speech with MBG	50	43	86%	5	89.6%
SFX with MBG	40	35	87.5%	6	85.4%
Harmonic SFX	40	36	90%	0	100%
Non-harmonic SFX	600	591	98.5%	29	95.3%

1.3 SEGMENTATION AND INDEXING RESULTS

We tested the segmentation procedure with audio clips recorded from movies and video programs. With Pentium333 PC/Windows NT, segmentation and classification tasks can be achieved together with less than one eighth of the time required to play the audio clip. We made a demonstration program for on-line audiovisual data segmentation and

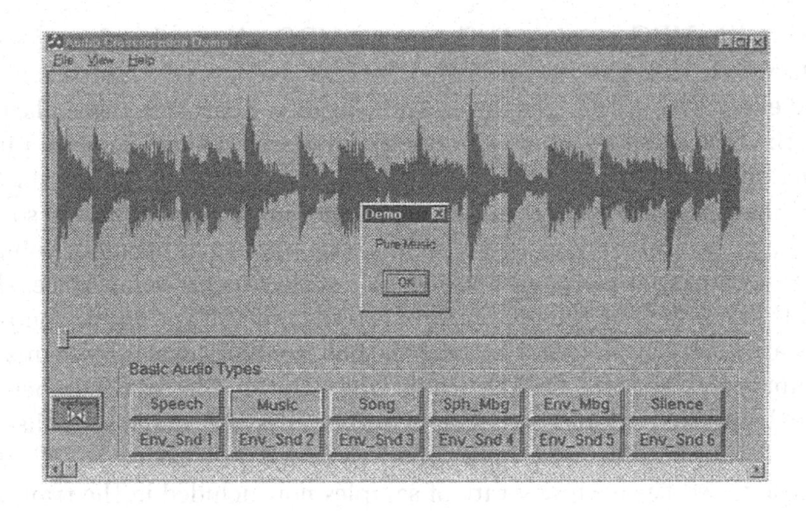

Figure 7.1. **Demonstration of generic audio data classification.**

indexing as shown in Figure 7.2, where different types of audio data are represented by different colors. Displayed on this figure is the segmentation and indexing result for a 42-second long audio clip recorded from a Spanish cartoon video called "Don Quijote de la Mancha". The first segment in this audio clip is song performed by children and with musical instrument accompanyment, which is indexed as "song". Then, after a period of silence which is indexed as "silence", there is a segment of female speech, and it is indexed as "pure speech". Afterwards, there is a short pause indexed as "silence" and followed by a segment of music which is indexed as "pure music". Next, with the music as background, comes the speech of an old male, and the segment is indexed as "speech with the music background". Finally, the music stops, and there is speech of a boy which is indexed as "pure speech".

For another example, an audio clip recorded from the movie "Washington Square" was segmented as illustrated in Figure 7.3. In this 50-second long audio clip, there is first a segment of speech spoken by a female (indexed as "pure speech"), then a segment of screams by a group of people (indexed as "non-harmonic and irregular environmental sound"), followed by a period of unrecognizable conversation of multiple people simultaneously mixed with baby cry (indexed as the mix of harmonic and non-harmonic sounds). Then, a low volume music appears as the background (indexed as "environmental sound with music background"). Afterwards, there is a segment of music with very low level environmental sounds as background (indexed as "pure music").

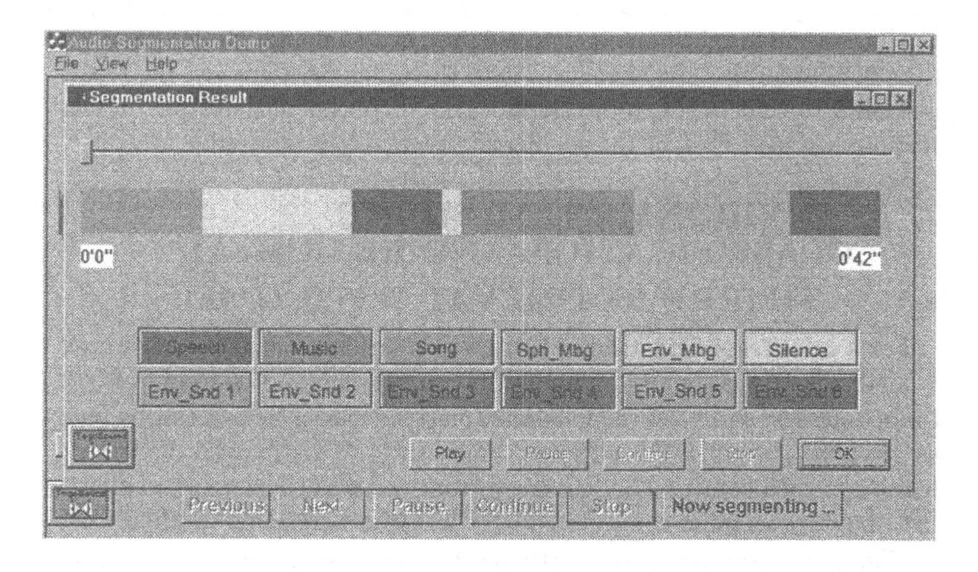

Figure 7.2. Demonstration of audiovisual data segmentation.

And finally, there is a short conversation between a male and a female (indexed as "pure speech").

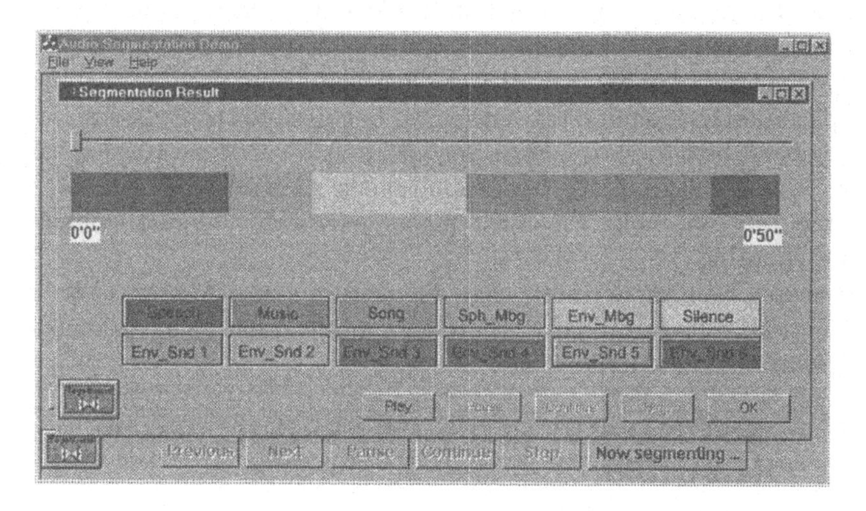

Figure 7.3. Segmentation of a movie audio clip.

Besides the above two examples, we also performed experiments on twenty or so other audio clips. In general, boundaries between segments of different audio types are set quite precisely with a precision within one second as compared to human perception. Using human judgement as the ground truth, our algorithm is sensitive enough to detect more than 95% of audio type changes. As to the indexing accuracy, the result is similar to that of the audio classification experiment described in last section, i.e. over 90% of the segments are correctly indexed.

2. ENVIRONMENTAL SOUND CLASSIFICATION AND RETRIEVAL

2.1 TIMBRE RETRIEVAL WITH GMM

For a brief test of the timbre feature extraction scheme, we organized a small set of data which includes 100 simple-timbred sounds from 12 timbre classes, and used the Gaussian mixture model (which can be seen as a special case of HMM with one single state) to distinguish the timbre feature of each sound. The feature vector used was the smoothed short-time power spectrum computed with the "autocorrelation method".

The GMM parameters were trained for every sound in the set. Then, one piece of the "applause" sound was chosen as the query sound, and its feature vector sequence was matched to each of the GMMs. The resulted top ten sounds in the likelihood ranking list, which were considered most similar to the input query, were from the following classes: no.1-7: applause; no.8-9: river-flow; no.10: applause. This result is reasonable, because the applause by a crowd of people and the fast, big flow of river sometimes really sound alike.

For another example, a sound of "dog bark" was used as the query sample, and the top ten retrieved sounds belonged to these classes: no.1-5: dog bark; no.6-7: shout; no.8: dog bark; no.9: shout; no.10: dog bark. There were 7 pieces of dog bark sound in the data set which all entered the top ten list, while the other places were taken by the sounds of crowd-shouting.

It is seen from these results that, the feature vectors extracted for denoting timbre character work effectively in differentiating environmental sounds of different timbre classes.

2.2 SOUND EFFECTS CLASSIFICATION RESULTS

We collected a set of 210 environmental sound clips of 18 semantic classes from the audio database, with 6-18 sound clips in each class. This set was divided into two subsets: the training set and the testing set. There are 110 samples in the training set, with 4-10 samples from each class; and there are 100 samples in the testing set, with 2-8 samples from each class. An HMM parameter set was trained for each class with the samples in the training set. Then, the sound clips in both the training set and the testing set were classified, respectively, by matching with the HMM parameters of each class. The highest likelihood indicated the class to which the sound should be indexed. As shown in Table 7.3 (where "sample #" refers to the number of samples in the set, and "correct sample #" is the number of samples which were correctly indexed in the set), all the sound clips in the training set were correctly classified. For the testing set, 86 out of 100 were correctly indexed, suggesting an accuracy rate of 86%.

Table 7.3. Results of sound effects classification using HMM.

Class Name	Training Set		Testing Set	
	sample #	correct sample #	sample #	correct sample #
Applause	8	8	8	8
Dog-bark	6	6	6	6
Bird-cry	10	10	8	8
Crowd	4	4	4	3
Explosion	6	6	6	5
Foot-step	4	4	4	4
Glass-break	4	4	2	2
Gun-shoot	6	6	6	5
Knock	4	4	4	2
Laughter	8	8	6	5
Plane-take off	6	6	6	4
Rain	6	6	6	5
River-flow	8	8	6	5
Touch-tone	6	6	6	4
Thunder	8	8	6	6
Water-pour	4	4	4	4
Whistle	6	6	6	6
Windstorm	6	6	6	4
Total	110	110	100	86

Most misclassifications occurred within those semantic classes that have similar sounds, such as among explosion, thunder and gun shot, among rain, river and applause, between touch-tone and whistle, between plane taking off and windstorm, and between crowd and laughter.

However, there are two special cases to be mentioned. One is the classification for quasi-periodic sounds, such as those in the "Knock" class. Sounds of this kind are recognized to a large extent by rhythm, by which we mean the change pattern of energy, rather than by the spectral envelope. For example, the sounds of knock may have quite different spectral shapes depending on how and what you knock (e.g. with palm or finger, on wooden door or glass), but should have very similar rhythms. The other case is the comparison of harmonic sounds. The feature of timbre for non-harmonic sound is the spectral envelope. While for harmonic sound, timbre is mostly represented by the harmonic lines. One example may be the sounds in the class "Touch-tone". Although we can recognize them all as touch-tone sounds by human perception, as the positions of their harmonic lines are different in the frequency domain, they were not considered all as similar by just comparing the spectral envelopes. Quasi-periodic and harmonic environmental sounds are two basic types of audio which are detected in the coarse-level classification, and special recognition schemes for them are to be worked on.

Another issue to be mentioned is that, as the sounds within one semantic class may sound quite differently (e.g. the sounds of applause by one person and that by a large crowd of people), and there may be sounds that sound very alike but belong to different semantic classes (such as "Explosion", "Thunder", and "Gun-shot"), a dynamic HMM merging and splitting mechanism should be built to adapt to such cases.

2.3 SOUND EFFECTS RETRIEVAL RESULTS

Using the sound sample set mentioned above for testing audio retrieval performances, we trained HMM parameter set for each of the 210 sound clips. Four clips were taken to be the query sound, which were an applause sound, a bird-cry sound, a thunder sound, and a whistle sound. The classes to which the top ten retrieved sounds (which are regarded as the most similar sounds to the input query in the sample set) belong for each query sound are listed in Table 7.4.

Several conclusions may be summarized from the retrieval results. First, for each of the query sounds, at least five of the top ten retrieved sounds, as well as all of the top three retrieved sounds were from the class to which the query sound belonged. Second, most sounds in the class to which the query sample belonged had a rather high likelihood even though some of them might not enter the top ten list. Third, sounds which ranked high in the likelihood list normally were from the same several classes (note this is also true with the results of timbre retrieval using GMM).

Table 7.4. Results of sound effects retrieval based on HMM.

Query Sound		Applause	Bird-cry	Thunder	Whistle
Retrieved Sounds	No.1	Applause	Bird-cry	Thunder	Whistle
	No.2	Applause	Bird-cry	Thunder	Whistle
	No.3	Applause	Bird-cry	Thunder	Whistle
	No.4	River-flow	Water-pour	Thunder	Whistle
	No.5	Applause	Laughter	Thunder	Whistle
	No.6	Gun-shoot	Water-pour	Dog-bark	Whistle
	No.7	Applause	Water-pour	Thunder	Whistle
	No.8	Applause	Bird-cry	Thunder	Whistle
	No.9	Applause	Water-pour	Thunder	Laughter
	No.10	Rain	Bird-cry	Dog-bark	Laughter

Thus, a classification-based retrieval scheme may work as follows. For each sound in the top ten list, the user is asked "want more like this?". He may listen to one retrieved sound, and then choose to browse or discard other sounds in the class to which the retrieved sound belongs. In this way, the user may get the whole class of sounds which he wants, and avoid the classes of sounds that he does not want. As the high-ranked sounds come from only several semantic classes, the selection efficiency is high. Also, the sound clips in the class which is selected may be ordered by their likelihood to the query sound, and are presented to the user by this order.

3. SHOT CHANGE DETECTION AND KEYFRAME EXTRACTION

3.1 SHOT CHANGE DETECTION RESULTS

The proposed scheme for shot change detection was tested on five video clips, which were extracted from a newscast video, a documentary video about culture, a feature movie, a scientific video, and a cartoon video, respectively. Each video clip is around 4-minute long. Various kinds of shot changes and camera motions included in these video clips are listed in Table 7.5. There are also many shots involved which have appearance/disappearance of captions, luminance changes (including photoflashes), fast object motions, and noise (such as local flashes) that may cause relatively large values of the histogram difference.

Table 7.5. Shot changes and camera motions in test video clips.

Video clip		News video	Culture video	Movie	Science video	Cartoon	Total
Abrupt change		27	17	19	12	29	104
Gradual transition	Dissolve	3	36	0	3	15	57
	Wipe	0	2	0	0	0	2
	Fade-in	0	0	0	0	1	1
	Fade-out	0	0	1	0	2	3
Total shot changes		30	55	20	15	47	167
Camera motion	Panning	6	16	12	3	9	46
	Tilting	1	3	0	1	0	5
	Zoom-in	1	1	0	1	7	10
	Zoom-out	0	1	0	1	4	6
Total camera motions		8	21	12	6	20	67

Detection results are displayed in Table 7.6. Compared to the performance of the modified twin-comparison approach as shown in Table 6.1, significant improvements are achieved by combining information in the Y- and V- components. Only three gradual shot transitions were missed. All of them are dissolves occurring simultaneously with fast camera panning which makes it difficult to differentiate. Reasons for the ten false alarms include: large size object movement together with fast panning or luminance change, appearance or moving of the caption together with panning, fast panning which blurs the frames, images of flame having irregular luminance change, slow tilting not detected. Large values of the histogram difference arise in these cases. Since these situations rarely occur, the overall reliability of the proposed scheme is still high.

3.2 KEYFRAME EXTRACTION RESULTS

The adaptive keyframe extraction method proposed in this work was experimented on a dozen of shots selected from test video clips. Satisfactory results were obtained in almost all cases. Two examples are shown in Figures 7.4 and 7.5, respectively. In both cases, the threshold was set to 200 for the cumulative difference value of the Y-component histogram to detect a new keyframe from the shot, and four keyframes were generated from each shot. The first shot was picked from a documentary

Table 7.6. Shot change detection results of the proposed scheme.

Testing video clips	Actual Number of shot changes	Correct detects		False alarms	
		Number	Sensitivity	Number	Recall
News video	30	30	100%	1	97%
Cultural video	55	52	95%	4	93%
Movie	20	20	100%	1	95%
Scientific video	15	15	100%	1	94%
Cartoon	47	47	100%	3	94%
Total	167	164	98%	10	94%

video. It contains 284 frames, and has a large scale zooming-out motion of the camera. As shown in Figure 7.4, the zooming procedure is properly represented by the selected four keyframes. The primary objects (the person in the first keyframe whose voice is the offscene sound of this shot and the full view of the building) are well presented. The second one is a long shot with 1023 frames in the beginning part of a cartoon video. It has at first a panning motion, and then a zooming-in motion. There are also captions on some frames. Again, dominant objects and camera motions of the shot are reflected in extracted keyframes. People can have a quick grasp of the content within this 41-second shot by only looking at these four frames.

In another attempt, we extracted keyframes from one shot in a newscast video. The shot contains 531 frames and fast camera panning motions. The threshold was set to 100 and the maximum number of keyframes allowed for one shot was chosen to be 4. Sixteen frames were first selected through the thresholding process, and then four keyframes were extracted from these candidates as shown in Figure 7.6. Compared to the use of the thresholding process alone (e.g. three keyframes were obtained when the threshold was set to 200), this two-step procedure was shown to produce keyframes which better represent those shots with complex content.

4. INDEX TABLE GENERATION

4.1 INDEX TABLE FOR NEWS BULLETIN

We built content models and generated index tables for two types of video, i.e. news bulletin and documentary. The index table of one newscast video package has a structure containing three levels. The

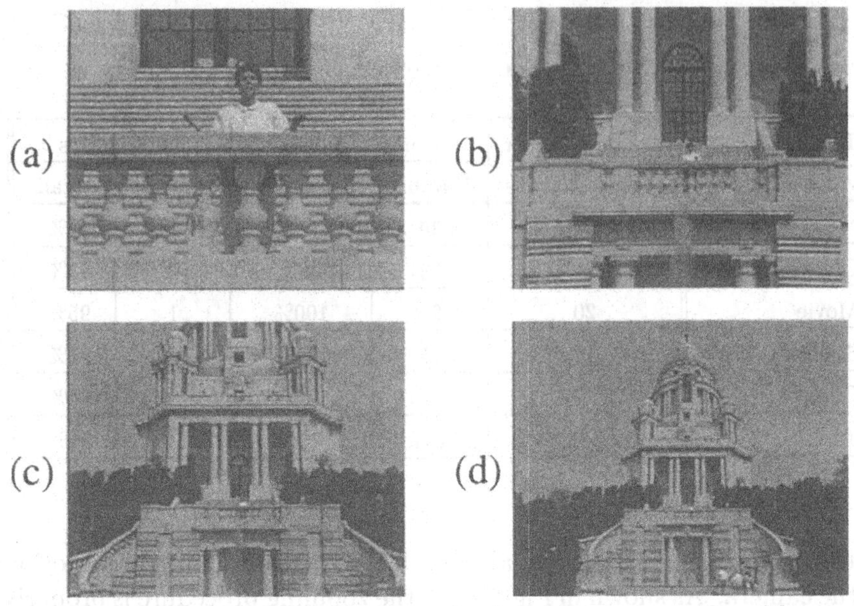

Figure 7.4. **Keyframes extracted from one shot in a documentary video with camera zoom-out motion.**

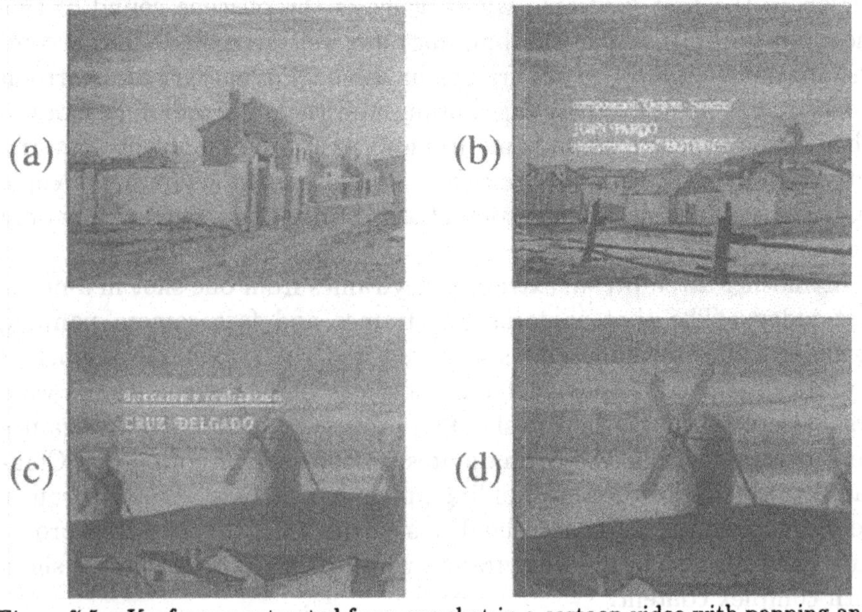

Figure 7.5. **Keyframes extracted from one shot in a cartoon video with panning and zoom-in motions.**

first level contains information about the whole video source which has a similar format among different video types. The second level is for the description of news items. The time interval and the first 10-second speech of the anchorperson of each news item are ordered chronologically

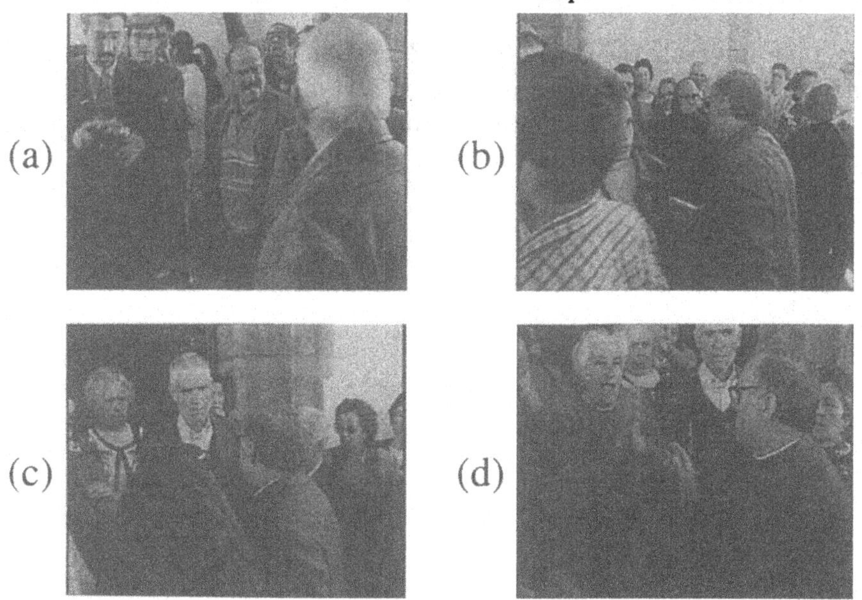

Figure 7.6. Keyframes extracted from one shot in a newscast video with fast camera panning motion.

in this level. The news items are separated from each other by detecting shots of the anchorperson with color histogram comparison and pauses in anchorperson's speech. Then, the third level description is the content within each news item, including time intervals and indices of audio scenes, as well as time intervals and keyframes of shots in the news item. The audio and visual indices are interleaved in each news item according to the temporal order.

4.2 INDEX TABLE FOR DOCUMENTARY

For documentary video, there are also three levels of structure in the index table: package, scene and shot. The package-level metadata include facts about the video program such as length, frame rate and data format. A scene break is defined only when both audio scene change and visual shot change occur. However, the definition of audio scene change is slightly different from the segmentation rule as described in Chapter 4. For example, with the same music episode continuing, appearance and disappearance of speech on top of music is regarded as the break between two audio types (i.e. "pure music" and "speech with music background"), but not an audio scene break. It will be viewed within one scene as long as the same music continues. Nevertheless, if there is change of music (i.e. two different episodes) or long pause between two speech segments, an audio scene break will be claimed. For each scene, the time interval, the first 10-second offscene speech (if there is any) and the first image frame are included at the second level description. Then, for each shot, there are the time interval, the audio type index and keyframes contained in the third level description.

VI

CONCLUSION

Chapter 8

CONCLUSION AND EXTENSIONS

1. CONCLUSION

A system was proposed in this thesis for automatic segmentation, indexing and retrieval of audiovisual data based on multimodal media content analysis. The video stream was demultiplexed into different media types such as audio, image and caption. An index table was generated for each video clip by combining results from content analysis of these diverse media types. Structures for different video types were described, and models were built for each video type individually. This general modeling and structuring of video content parsing is very unique. It achieves more functions than existing approaches which normally adopt a single model with focus on the pictorial information alone.

For content-based management of audiovisual data, a hierarchical system consisting of three stages was developed . In the first stage, the task of on-line segmentation and classification of accompanying audio signals into twelve basic types of sound was accomplished. The boundaries were precisely set, and an accurate classification rate higher than 90% was achieved. The procedure is generic and model free. In the second stage, fine-level classification of environmental sounds by using the hidden Markov model was performed. Experimental results showed that an accuracy rate of 86% was obtained. Finally, based on the classification approach, a query-by-example retrieval scheme for sound effects was proposed and proved to be very effective.

For content analysis of image sequences, an efficient and robust method was developed for shot change detection. This new method was derived from the twin-comparison algorithm with a new ingredient, i.e. the histogram difference of the Y- and V- components was incorporated. It was shown that the proposed method achieved both the sensitivity rate and the recall rate at around 95% with various kinds of test video. A scheme was also proposed for adaptive keyframe extraction based on histogram comparison. It was demonstrated by experiments that it could generate keyframes that properly represent the content of a shot.

Further study will be performed on manipulating audiovisual data in the compression domain. Work on system integration/applications and continual effort in the MPEG-7 standard will be pursued. These research directions will be described in the following sections.

2. FEATURE EXTRACTION IN THE COMPRESSION DOMAIN

As most digital audiovisual data available these days are in the MPEG format, we will work on the extraction of audio and visual features directly from the MPEG coded bit stream. The block diagram of a typical MPEG coding system is shown in Figure 8.1, which contains the system layer, audio and video compression modules [59]. The main function of the MPEG system layer is to combine one or more audio and video compressed bit streams into a single bit stream. It defines the data stream syntax for timing control and the interleaving and synchronization of audio and video bit streams [60]. The MPEG audio codec defines three layers of coding schemes of increasing complexity to achieve better subjective quality. MPEG video is specifically designed for compression of image sequences. Except for the special case of a scene change, pictures in an image sequence tend to be quite similar from one to the next. The MPEG compression scheme takes advantage of this similarity. It consists of two main coding modes: namely, interframe and intraframe coding. Compression techniques that exploit information from certain reference pictures in the sequence are usually called interframe techniques. When a scene change occurs, interframe compression does not work well. Compression techniques that only use information from a single picture are usually called intraframe techniques.

In this research, audio features such as short-time energy function, average zero-crossing rate, fundamental frequency, and spectral peak track were extracted from raw audio data. For shot change detection and keyframe extraction, we used features of color histograms (of the YUV format) and motion vectors. It is interesting to investigate methods to estimate these features from the MPEG bit stream or the subband/transformed data. For example, spectral peak tracks can be detected from only certain frequency bands so that interference from other frequency bands may be avoided. Color histograms (for the YUV format) can also be obtained from the transformed domain [35]. While

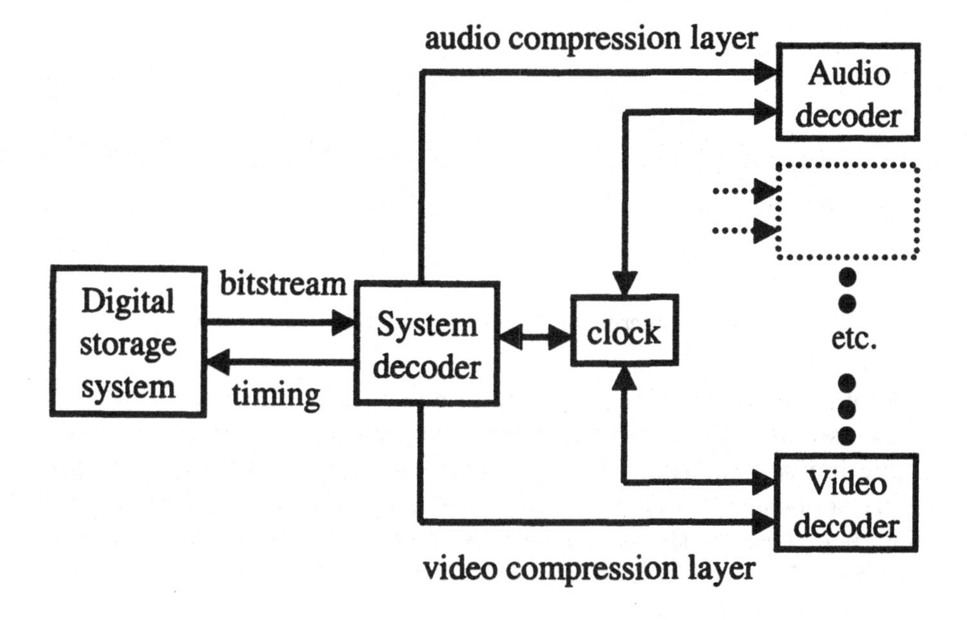

Figure 8.1. The MPEG system structure.

estimation of motion vectors is a rather time-consuming work in the raw data domain, their values are provided in the compressed domain (obtained in the encoding stage) which can be used directly for visual content analysis. Information of shot changes and I-frames is also included in the compressed data stream.

3. SYSTEM INTEGRATION AND APPLICATIONS

Techniques developed in this research for audiovisual data modeling, segmentation, indexing and retrieval, as well as tools for audio and visual content analysis can be integrated for particular applications. There are several possibilities.

One is to build an information filtering and retrieval engine for multicast backbone video (including video conferencing and broadcast video over the Internet). Algorithms studied in this research can be potentially used to produce semantic annotation of multicast video. Audiovisual data and their corresponding metadata can be synchronized and transmitted over the multicast backbone. The filtering and retrieval engine selects proper parts and/or channels of video data by matching users' interest with the metadata.

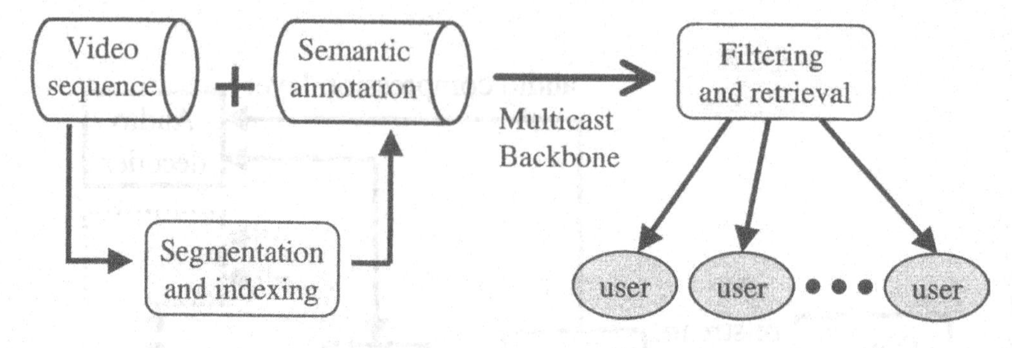

Figure 8.2. Information filtering and retrieval of multicast backbone video.

Another possibility is to develop an audiovisual surveillance system. Cameras and audio recording devices may be installed in places with high criminal activities. The developed algorithms can be used to recognize sounds (e.g. explosion, glass-breaking, gun-shot, and cries) and shots (e.g. those with fast object movements) associated with criminal activities. An action will be triggered automatically if some event occurs as shown in Figure 8.3.

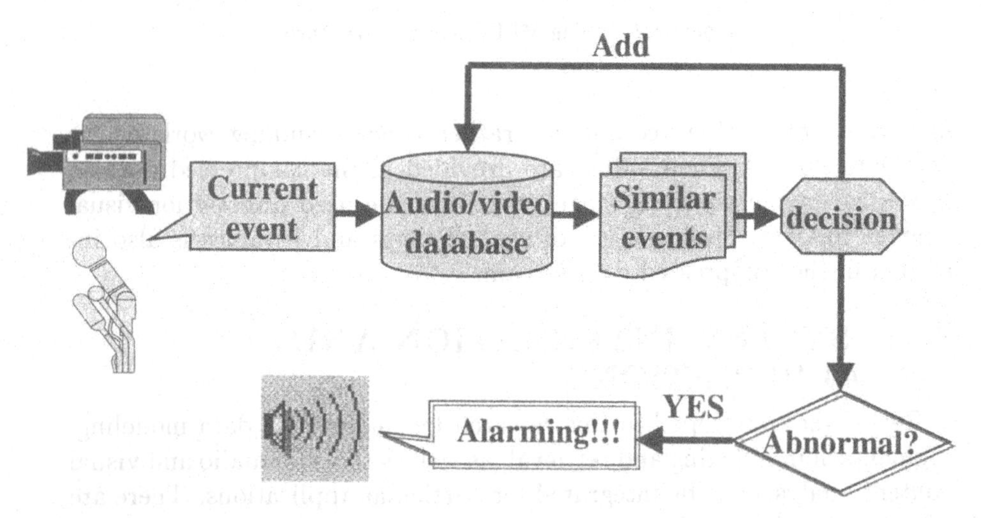

Figure 8.3. The use of audiovisual data indexing and retrieval techniques in surveillance applications.

4. CONTRIBUTIONS TO MPEG-7

The normative part of the MPEG-7 Standard includes description schemes, descriptors, the description definition language, and coding methods of descriptions. Extraction methods, search methods, and evaluation and validation techniques will be developed in the non-normative part of MPEG-7. With the MPEG-7 terminology [7], a descriptor(D) defines the syntax and semantics of a representation entity for a feature. For example, a time-code may be the descriptor of duration, while both color moments and histograms may be descriptors for the color feature. A description scheme(DS) consists of one or more descriptors and description schemes. DS specifies the structure and semantics of relationships between them. A description of multimedia material in MPEG-7 contains a fully or partially instantiated DS. Description schemes are specified in the description definition language (DDL). DDL allows creation of new description schemes and descriptors and extension of existing description schemes. Relations of these terms are illustrated in Figure 8.4.

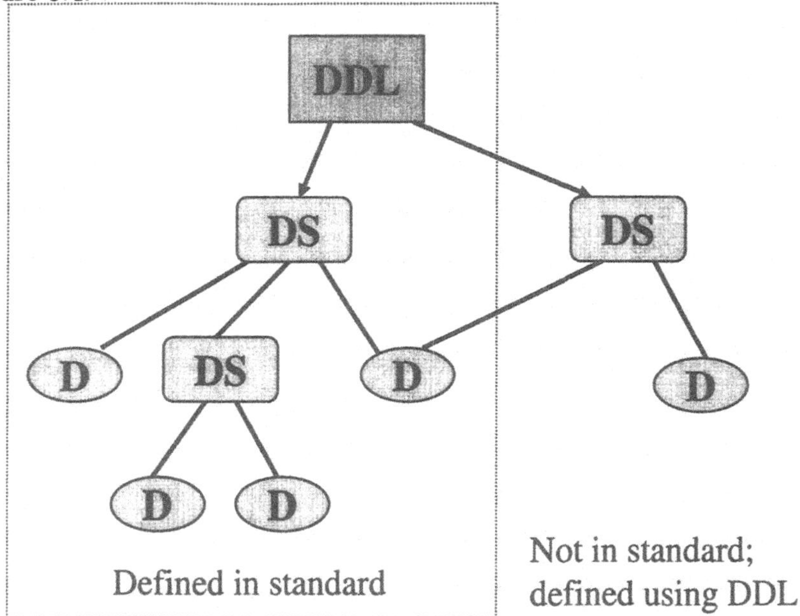

Figure 8.4. Descriptors(D), Description Schemes(DS), and the Description Definition Language(DDL) in MPEG-7.

We submitted proposals to the Audio Part of MPEG-7 containing description schemes and descriptors for generic audio data segmentation and indexing, as well as for classification and retrieval of environmental sounds. We also submitted extraction and search methods for the descriptor computation, audio classification, and similarity measuring procedures. Most of the proposals were recommended to enter the Core Experiments (CE) of MPEG-7 at the evaluation meeting held in February 1999. These descriptions and processing methods will find applications in domains such as storage and retrieval of audio/video databases, delivery of audio/video segments for professional media production, sound effects libraries, movie scene retrieval by memorable auditory events, user agent driven media selection and filtering, semi-automated multimedia editing, surveillance, and so on. We will continue to conduct core experiments for proposed descriptors and description schemes. In the meanwhile, we plan to contribute the proposed video content parsing method as a multimodal description scheme to the Video Part of the MPEG-7 Standard.

References

[1] S. W. Smoliar and H. Zhang, "Content-based video indexing and retrieval," *IEEE Multimedia*, pp. 62-72, Summer 1994.

[2] M. Flickner, H. Sawhney, W. Niblack, *et al.*, "Query by image and video content: the QBIC system," *Computer*, vol. 28, no. 9, pp. 23-32, 1995.

[3] "Final text of ISO/IEC FCD 14496-3 Audio," *ISO/IEC JTC 1/SC 29/WG 11 Document N2203*, Tokyo, Mar. 1998.

[4] A. Gersho, "Advances in speech and audio compression," *Proceedings of the IEEE*, vol. 82, no. 6, pp. 900-918, 1994.

[5] D. K. Freeman, G. Cosier, C. B. Southcott, *et al.*, "The voice activity detector for the Pan-European digital cellular mobile telephone service," *Proceedings of ICASSP'89*, vol. 1, pp. 369-372, 1989.

[6] H. J. Zhang, A. Kankanhalli, and S. W. Smoliar, "Automatic partitioning of full-motion video," *Multimedia Systems*, vol. 1, no. 1, pp. 10-28, 1993.

[7] "MPEG-7 Requirements Document," *ISO/IEC JTC1/SC29/WG11 Document N2461*, Atlantic City, Oct. 1998.

[8] "MPEG-7 Applications Document," *ISO/IEC JTC1/SC29/WG11 Document N2462*, Atlantic City, Oct. 1998.

[9] S.-F. Chang, W. Chen, H. J. Meng, *et al.*, "A fully automated content based video search engine supporting spatio-temporal queries," *IEEE Transactions on Circuits and Systems for Video Technology*, vol. 8, no. 5, pp. 602-615, 1998.

[10] J. Huang, Z. Liu, and Y. Wang, "Integration of audio and visual information for content-based video segmentation," *Proceedings of IEEE Conference on Image Processing*, Chicago, Oct. 1998.

[11] M. R. Naphade, T. Kristjansson, B. Frey, *et al.*, "Probabilistic multimedia objects (MULTIJECTS): a novel approach to video indexing and retrieval in multimedia systems," *Proceedings of IEEE Conference on Image Processing*, Chicago, Oct. 1998.

[12] J. S. Boreczky and L. D. Wilcox, "A hidden Markov model framework for video segmentation using audio and image features," *Proceedings of ICASSP'98*, pp. 3741-3744, Seattle, May 1998.

[13] C.-S. Li, R. Mohan, and J. R. Smith, "Multimedia content description in the InfoPyramid," *Proceedings of ICASSP'98 Special Session on Signal Processing in Modern Multimedia Standards*, Seattle, May 1998.

[14] M. F. Vetter, "Dynamic metadata dictionary structure," *SMPTE Engineering Working Group on Metadata*, Tewksbury, MA, Aug. 1998.

[15] J. Saunders, "Real-time discrimination of broadcast speech/music," *Proceedings of ICASSP'96*, vol. II, pp. 993-996, Atlanta, May 1996.

[16] E. Scheirer and M. Slaney, "Construction and evaluation of a robust multifeature speech/music discriminator," *Proceedings of ICASSP'97*, Munich, Germany, April 1997.

[17] L. Wyse and S. Smoliar, "Toward content-based audio indexing and retrieval and a new speaker discrimination technique," downloaded from *http://www.iss.nus.sg/People/lwyse/lwyse.html*, Institute of Systems Science, National Univ. of Singapore, Dec. 1995.

[18] D. Kimber and L. Wilcox, "Acoustic segmentation for audio browsers," *Proceedings of Interface Conference*, Sydney, Australia, July 1996.

[19] S. Pfeiffer, S. Fischer, and W. Effelsberg, "Automatic audio content analysis," downloaded from *http://www.informatik.uni-mannheim.de/ pfeiffer/publications/*, Praktische Informatik IV, University of Mannheim, Germany, Apr. 1996.

[20] A. Ghias, J. Logan, and D. Chamberlin, "Query by humming - musical information retrieval in an audio database," *Proceedings of ACM Multimedia Conference*, pp.231-235, Anaheim, CA, 1995.

[21] J. Foote, "Content-based retrieval of music and audio," *Proceedings of SPIE'97*, Dallas, 1997.

[22] E. Wold, T. Blum, D. Keislar, *et al.*, "Content-based classification, search, and retrieval of audio," *IEEE Multimedia*, pp. 27-36, Fall 1996.

[23] G. Smith, H. Murase, and K. Kashino, "Quick audio retrieval using active search," *Proceedings of ICASSP'98*, pp. 3777-3780, Seattle, May 1998.

[24] Z. Liu, J. Huang, Y. Wang, *et al.*, "Audio feature extraction and analysis for scene classification," *Proceedings of IEEE 1st Multimedia Workshop*, 1997.

[25] Z. Liu, J. Huang, and Y. Wang, "Classification of TV programs based on audio information using hidden Markov model," *Proceedings of IEEE Second Workshop on Multimedia Signal Processing*, pp. 27-32, Redondo Beach, CA, Dec. 1998.

[26] Z. Liu and Q. Huang, "Classification of audio events in broadcast news," *Proceedings of IEEE Second Workshop on Multimedia Signal Processing*, pp. 364-369, Redondo Beach, CA, Dec. 1998.

[27] N. Patel and I. Sethi, "Audio characterization for video indexing," *Proceedings of SPIE Conference on Storage and Retrieval for Still Image and Video Databases*, vol. 2670, pp. 373-384, San Jose, 1996.

[28] K. Minami, A. Akutsu, H. Hamada, *et al.*, "Video handling with music and speech detection," *IEEE Multimedia*, pp. 17-25, Fall 1998.

[29] A. S. Bregman, *Auditory scene analysis: the perceptual organization of sound*, Cambridge, Mass.: MIT Press, 1990.

[30] G. J. Brown and M. Cooke, "Computational auditory scene analysis," *Computer Speech and Language*, vol. 8, no. 2, pp. 297-336, 1994.

[31] M. Weintraub, *A Theory and Computational Model of Auditory Monaural Sound Separation*, PhD thesis, Standford University, Dept. of Electrical Engineering, Palo Alto, CA, 1985.

[32] D. P. W. Ellis, *Prediction-Driven Computational Auditory Scene Analysis*, PhD thesis, MIT, Dept. of Electrical Engineering and Computer Science, Cambridge, MA, 1996.

[33] B. L. Vercoe, W. G. Gardner, and E. D. Scheirer, "Structured audio: creation, transmission, and rendering of parametric sound representations," *Proceedings of the IEEE*, vol. 86, no. 5, pp. 922-939, May 1998.

[34] G. Ahanger and T. D. C. Little, "A survey of technologies for parsing and indexing digital video," *Journal of Visual Communication and Image Representation*, vol. 7, no. 1, pp. 28-43, 1996.

[35] J. Meng, Y. Juan, and S. F. Chang, "Scene change detection in a MPEG compressed video sequence," *SPIE Symposium on Electronic Image: Science and Technology - Digital Video Compression: Algorithms and Technologies*, SPIE vol. 2419, San Jose, Feb. 1995.

[36] H. J. Zhang and S. W. Smoliar, "Developing power tools for video indexing and retrieval," *Proceedings of IS&T/SPIE Conference on Storage and Retrieval for Image and Video Databases II*, SPIE vol. 2185, pp. 140-149, San Jose, Feb. 1994.

[37] D. Zhong, H. J. Zhang, and S.-F. Chang, "Clustering methods for video browsing and annotation," *Proceedings of SPIE Conference on Storage and Retrieval for Image and Video Database*, San Jose, Feb. 1996.

[38] L. Rabiner and R. Schafer, *Digital Processing of Speech Signals*, Prentice-Hall, Inc., New Jersey, 1978.

[39] A. Choi, "Real-time fundamental frequency estimation by least-square fitting," *IEEE Transactions on Speech and Audio Processing*, vol. 5, no. 2, pp. 201-205, Mar. 1997.

[40] B. Doval and X. Rodet, "Estimation of fundamental frequency of music sound signals," *Proceedings of ICASSP'91*, vol. 5, pp. 3657-3660, Toronto, Apr. 1991.

[41] W. B. Kuhn, "A real-time pitch recognition algorithm for music applications," *Computer Music Journal*, vol. 14, no. 3, pp. 60-71, Fall 1990.

[42] F. Everest, *The Master Handbook of Acoustics*, McGraw-Hill, Inc., 1994.

[43] S. Haykin, *Adaptive Filter Theory*, Prentice Hall, 1991.

[44] E. Miyasaka, "Timbre of complex tone bursts with time varying spectral envelope," *Proceedings of ICASSP'82*, vol. 3, pp. 1462-1465, Paris, May 1982.

[45] M. Leman, *Music, Gestalt, and Computing: Studies in Cognitive and Systematic Musicology*, Springer, 1997.

[46] K. C. Pohlmann, *Principles of Digital Audio*, McGraw-Hill, Inc., 1995.

[47] F. Winckel, *Music, Sound and Sensation*, Dover Publications, Inc., New York, 1967.

[48] H. Benade, *Fundamentals of Musical Acoustics*, Dover Publications, Inc., New York, 1976.

[49] M. Hawley, *Structure Out of Sound*, PhD thesis, MIT, Dept. Media Arts and Sciences, Cambridge, Mass., 1993.

[50] L. Rabiner and B. Juang, *Fundamentals of Speech Recognition*, Prentice-Hall, Inc., New Jersey, 1993.

[51] J. R. Deller, J. G. Proakis, and J. H. L. Hansen, *Discrete-Time Processing of Speech Signals*, Macmillan Publishing Company, New York, 1993.

[52] G. E. Pelton, *Voice Processing*, McGraw-Hill, Inc., 1993.

[53] D. Reynolds and R. Rose, "Robust text-independent speaker identification using Gaussian mixture speaker models," *IEEE Transactions on Speech and Audio Processing*, vol. 3, no. 1, pp. 72-83, 1995.

[54] H. Gish and M. Schmidt, "Text-independent speaker identification," *IEEE Signal Processing Magazine*, pp. 18-32, Oct. 1994.

[55] S. Bow, *Pattern Recognition*, Marcel Dekker, Inc., 1984.

[56] A. Gersho and R. M. Gray, *Vector Quantization and Signal Compression*, Kluwer Academic Publishers, 1992.

[57] W. K. Pratt, *Digital Image Processing*, John Wiley & Sons, Inc., 1991.

[58] A. Nagasaka and Y. Tanaka, "Automatic video indexing and full-video search for object appearances," *Proceedings of 2nd Working Conference on Visual Database Systems*, pp. 119-133, 1991.

[59] J. L. Mitchell, W. B. Pennebaker, C. E. Fogg, and D. J. LeGall, *MPEG Video Compression Standard*, Chapman & Hall, New York, 1996.

[60] International Organization for Standardization document: *ISO/IEC 13818-1* (MPEG-2 Systems), March, 1994.

Index